DESIGNING WITH
FIELD-EFFECT
TRANSISTORS

Other McGraw-Hill Reference Books of Interest

Handbooks

Avallone and Baumeister • Marks' Standard Handbook for Mechanical Engineers
Benson • Audio Engineering Handbook
Benson • Television Engineering Handbook
Coombs • Printed Circuits Handbook
Coombs • Basic Electronic Instrument Handbook
Croft and Summers • American Electricians' Handbook
Fink and Beaty • Standard Handbook for Electrical Engineers
Fink and Christiansen • Electronic Engineers' Handbook
Harper • Handbook of Electronic Systems Design
Harper • Handbook of Thick Film Hybrid Microelectronics
Harper • Handbook of Wiring, Cabling, and Interconnecting for Electronics
Hicks • Standard Handbook of Engineering Calculations
Inglis • Electronic Communications Handbook
Juran and Gryna • Quality Control Handbook
Kaufman and Seidman • Handbook of Electronics Calculations
Kurtz • Handbook of Engineering Economics
Stout • Microprocessor Applications Handbook
Stout and Kaufman • Handbook of Microcircuit Design and Application
Stout and Kaufman • Handbook of Operational Amplifier Circuit Design
Tuma • Engineering Mathematics Handbook
Williams • Designer's Handbook of Integrated Circuits
Williams and Taylor • Electronic Filter Design Handbook

Other

Antognetti and Massobrio • Semiconductor Device Modeling with Spice
Antognetti • Power Integrated Circuits
Elliott • Integrated Circuits Fabrication Technology
Hecht • The Laser Guidebook
Mun • GaAs Integrated Circuits
Siliconix • Designing with Field-Effect Transistors
Sze • VLSI Technology
Tsui • LSI/VLSI Testability Design

DESIGNING WITH FIELD-EFFECT TRANSISTORS

Second Edition

SILICONIX INC.

Revised by

Ed Oxner

McGRAW-HILL PUBLISHING COMPANY

*New York   St. Louis   San Francisco   Auckland   Bogotá
Caracas   Hamburg   Lisbon   London   Madrid   Mexico
Milan   Montreal   New Delhi   Oklahoma City
Paris   San Juan   São Paulo   Singapore
Sydney   Tokyo   Toronto*

Library of Congress Cataloging-in-Publication Data

Designing with field-effect transistors / Siliconix Inc.—2nd ed. /
 rev. by Ed Oxner.
  p. cm.
 Includes bibliographies and index.
 ISBN 0-07-057537-1
 1. Field-effect transistors. 2. Transistor circuits.
 3. Electronic circuit design. I. Oxner, Edwin S. II. Siliconix
 Incorporated.
 TK7871.95.D47 1990
 621.381′5284—dc20   89-35590
                                                      CIP

1234567890 DOC/DOC 895432109

ISBN 0-07-057537-1

The editor for this book was Daniel A. Gonneau, the
designer was Elliot Epstein, and the production supervisor
was Dianne Walber. Project supervision by The Total Book.

Printed and bound by R. R. Donnelley & Sons Company.

*For more information about other McGraw-Hill materials, call
1-800-2-MCGRAW in the United States. In other countries, call
your nearest McGraw-Hill office.*

CONTENTS

v

CONTRIBUTORS

Written by the Applications Engineering Staff
of Siliconix Incorporated

Second Edition Revised by

Ed Oxner

Editor in Chief, First Edition

Arthur D. Evans

Contributors, First Edition

Arthur D. Evans
Walt Heinzer
Ed Oxner
Lee Shaeffer

PREFACE TO THE SECOND EDITION

In the nine years since the first edition of this book was published, the junction FET (JFET) has made little gain. However, such is not the case for double-diffused MOSFET (DMOS) technology. Both the small-signal DMOS and especially the large-signal or power [D]MOSFETs have made astounding advances and are likely, within the next decade, to become the principal players in new design, firmly but inexorably pushing aside the ubiquitous bipolar transistor.

Because of these advances in technology and application, the publisher decided to freshen up this popular and well-used book. Much of the original material concerning the junction FET remains—theory and applications are eternal. Keeping in mind that this book addresses small-signal FETs and applications, what is new in this edition is some detailed theory of DMOS, rounded out with applications involving both the small-signal DMOSFET as well as the lower power-rated power MOSFET.

With the increased interest in high-speed data transmission and processing, including advanced microprocessor and video applications, the double-diffused MOSFET (DMOS) will become a major player.

This book is designed principally for the FET user. The primary emphasis has been placed on allowing the reader to learn necessary information quickly, without becoming bogged down in complicated analyses. The first two chapters provide sufficient background information to answer many questions asked by FET circuit design engineers. The aim of these chapters is to give design engineers an intuitive sense of how to manipulate the circuits presented in the remaining chapters, which contain both theory and "cookbook" applications.

Some knowledge by the reader of semiconductor theory is assumed. A prior understanding of the concept of conduction by electrons and holes, of "forward" and "reverse" characteristics of a p-n junction, and of the "depletion" region at a p-n junction will be helpful in understanding FET theory.

Ed Oxner

PREFACE TO THE FIRST EDITION

The purpose of this book is to aid the electronics circuit designer in the utilization of the field-effect transistor (FET). Since its emergence from the development laboratory in about 1962, the FET has become an important and widely used component in the electronic industry.

With its importance increasing annually, both as a discrete component and in integrated circuits, it is essential that the serious circuit designer have an insight into how the FET behaves under different circuit and environmental conditions. It is also helpful to understand how its physical and electrical characteristics are interrelated. This book goes just deep enough into FET theory to provide that insight.

After the introductory chapter describing how the device works, the interrelationship of the various FET characteristics and how they relate to typical data sheet specifications is discussed.

Subsequent chapters deal with various categories of circuit applications of the FET. The order of these application chapters is rather arbitrary; each chapter stands on its own. Detailed circuit design examples are given to illustrate applications.

As this book was being readied for publication, the availability of FETs designed for power applications was expanding rapidly. The chapter on Power FETs spotlights this relatively new field.

Art Evans

1

FIELD-EFFECT TRANSISTORS: BASIC PRINCIPLES

1-1 INTRODUCTION

There are two general types of transistors: bipolar and unipolar. The unipolar, more commonly called the "field-effect transistor" (FET), is the subject of this book. The concept of controlling the electronic conduction in a solid by an electric field predates the invention of the bipolar transistor. J. E. Lilienfeld filed for a patent on such a device in 1925, as shown in Fig. 1-1. W. Shockley presented a comprehensive theory of the field-effect transistor in 1952. However, the commercial availability of the FET in the early 1960s followed the bipolar transistor by 8 or 10 years. The superior performance of the FET in many circuit applications previously utilizing bipolar transistors or vacuum tubes resulted in a rapid growth in its acceptance as an important electronic component.

Several types of semiconductor materials have been used for making FETs: silicon, germanium, gallium arsenide, and others. By far the most widely used is silicon, and unless otherwise specified all device types discussed in this book are silicon types.

The field-effect transistor (FET) is a class of electronic semiconductor device in which the conduction of a "channel" between source (S) and drain (D) terminals is controlled by an electric field impressed upon the channel via a gate (G) terminal. The conducting channel may utilize n-type carriers (electrons) or p-type carriers (holes). The electric field which controls the channel conduction may be introduced via a p-n junction (for a "junction" FET), a metal plate separated from the semiconductor channel by an oxide dielectric (for a metal-oxide-semiconductor FET), or a combination of the two methods. The polarity of the controlling electric field is a function of the type of carriers in the channel.

1

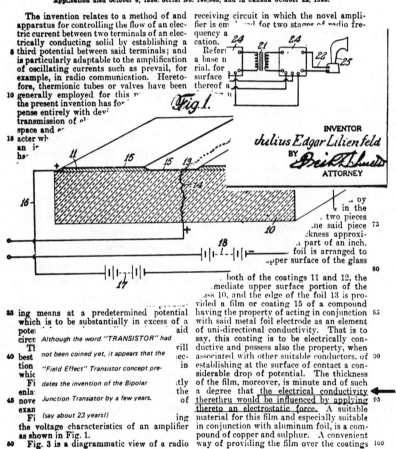

Patented Jan. 28, 1930 1,745,175

UNITED STATES PATENT OFFICE

JULIUS EDGAR LILIENFELD, OF BROOKLYN, NEW YORK

METHOD AND APPARATUS FOR CONTROLLING ELECTRIC CURRENTS

Application filed October 8, 1926. Serial No. 140,363, and in Canada October 22, 1925.

The invention relates to a method of and apparatus for controlling the flow of an electric current between two terminals of an electrically conducting solid by establishing a
5 third potential between said terminals; and is particularly adaptable to the amplification of oscillating currents such as prevail, for example, in radio communication. Heretofore, thermionic tubes or valves have been
10 generally employed for this p·
the present invention has fo·'
pense entirely with dev·
transmission of ·'
space and ·
15 acter wh·
an i·
h·

receiving circuit in which the novel amplifier is em   ·   ·-·' for two stages ·f ·adio frequency a
cation.
Refer·
a base n
rial, for
surface
thereof a
·· n

Fig. 1.

INVENTOR
Julius Edgar Lilienfeld
BY
ATTORNEY

·· by
·· in the
· two pieces
·ne said piece 75
·kness approxi-
·· part of an inch.
· foil is arranged to
·pper surface of the glass 80

16

+

18

10

25 ing means at a predetermined potential which is to be substantially in excess of a
pote·
circu

Although the word "TRANSISTOR" had
40 best
tion
whic
Fi
enla·
45 nove
exan
Fi
not been coined yet, it appears that the

"Field Effect" Transistor concept pre-

dates the invention of the Bipolar

Junction Transistor by a few years.

(say about 23 years!)
the voltage characteristics of an amplifier
50 as shown in Fig. 1.
Fig. 3 is a diagrammatic view of a radio

aid
·ill
·ec-
·iD
·tly
the
of

·ing

. both of the coatings 11 and 12, the
·mediate upper surface portion of the
·ass 10, and the edge of the foil 13 is provided a film or coating 15 of a compound 85
having the property of acting in conjunction
with said metal foil electrode as an element
of uni-directional conductivity. That is to
say, this coating is to be electrically conductive and possess also the property, when 90
associated with other suitable conductors, of
establishing at the surface of contact a considerable drop of potential. The thickness
of the film, moreover, is minute and of such
a degree that the electrical conductivity 95
therethru would be influenced by applying
thereto an electrostatic force. A suitable
material for this film and especially suitable
in conjunction with aluminum foil, is a compound of copper and sulphur. A convenient 100
way of providing the film over the coatings

FIGURE 1-1   "FET" patent predates the bipolar transistor patent by nearly 20 years!

1-2 FET SYMBOLS

The FET "family tree" shown in Fig. 1-2 indicates six different types. A different schematic symbol is utilized for each of these types. Figure 1-3 relates the symbols to the device types. Note that the direction of the arrow between the gate or body terminal and the channel will indicate the channel type. The "channel" of the enhancement types of MOS is broken, indicating that a channel must be "enhanced" to make the device operative.

These symbols will be utilized in the chapters that follow. Understanding these symbols will aid in understanding symbols for new device types that may be developed. Many device types are symmetrical with respect to source and drain; that is, source and drain functions may be interchanged. However, there are also devices which are not symmetrical. The symbol locates the gate electrode adjacent to the end of the device which is characterized by the manufacturer as the "source" terminal. (An attempt has been made to utilize symbols that correspond to IEEE/ANSI/ IEC and industry standards.)

As a word of caution, it should be pointed out that many schematics in technical literature neglect to show the "body" terminal of MOSFETs.

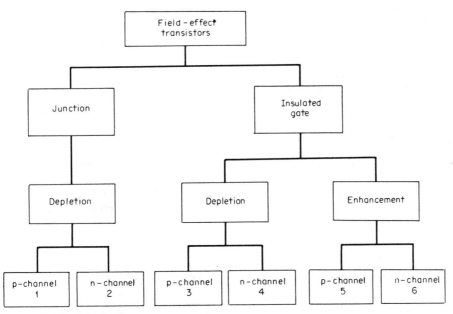

FIGURE 1-2 FET family tree.

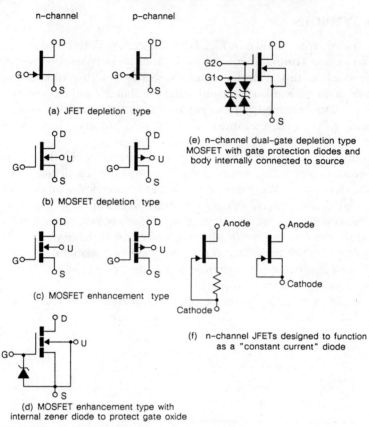

FIGURE 1-3   FET symbols.

In *most* cases it may be assumed that the body is internally connected to the source terminal; however, there are many cases in which the body is connected to some other potential, especially when the device is used as an analog switch. In integrated circuits, the body is often common to many MOSFETs within the same circuit.

1-3   PHYSICAL STRUCTURE OF SMALL-SIGNAL FETS

FET design and manufacturing details are beyond the scope of this book; however, a brief survey of some of the physical characteristics may be worthwhile. Manufacturing methods are similar to those utilized in other semiconductor devices such as integrated circuits, transistors, and diodes. Dimensions of the active areas, such as channel thickness, width, and length, may be controlled by a combination of epitaxial thickness, diffusion, ion implanting, etching, and photolithographic techniques.

1-3-1 The Junction FET

Figure 1-4 shows some of the steps involved in the manufacture of one type of n-channel junction field-effect transistor (JFET). The starting material in this case is a wafer of high-purity monocrystalline silicon which has been doped with an acceptor impurity (p-type). Onto this wafer a layer of silicon is epitaxially grown to a precisely controlled thickness and with a closely controlled donor (n-type) impurity concentration. A layer of silicon dioxide (glass) is formed on the wafer surface. By photolithographic techniques certain regions of this oxide are removed. The remaining oxide functions as a mask or barrier to p-type atoms, which are diffused into the unmasked regions. The masked region shown in Fig. 1-4c will become the source, drain, and channel region of the finished FET. A second oxidation, masking, and diffusion operation is used to create a top p-type gate which separates the source and drain. The depth of the top-gate diffusion and the initial thickness of the epitaxial layers determine the channel thickness. The width of the top gate stripe determines the channel length (source-to-drain dimension). Small variations in photographic dimensions, diffusion depths, epitaxial-layer thickness, and donor and acceptor impurity density will have an effect upon channel conduction, pinchoff voltage, gate-to-channel capacitance, and junction breakdown voltage. Lack of precise control of the physical parameters results in a production spread of the electrical characteristics of a given device type. In many instances a family of device types may be produced from a given set of photographic masks by varying the thickness of the epitaxial layer and the depth of the top-gate diffusion.

For a given channel thickness, width, and carrier concentrations, conduction will be an inverse function of channel length. For the structure described in Fig. 1-4 this length is controlled by the photographic process of etching the oxide window for top-gate diffusion, and by the sideways diffusion of the top gate. As the window size is decreased to make shorter, higher-conduction channels, control of the photographic process becomes more difficult. Because short channels are important for high-conduction and high-frequency devices, several methods for creating short-channel structures not so dependent upon very small photographic dimensions have been developed.

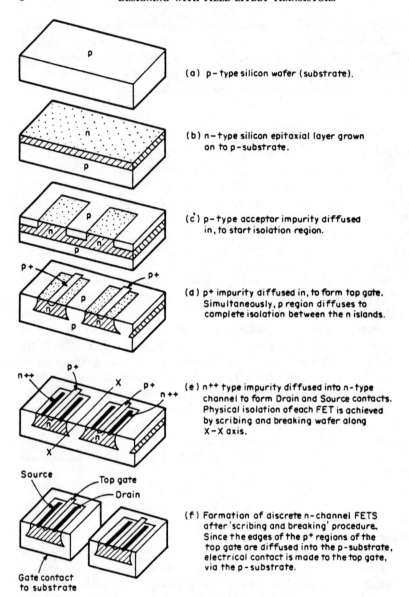

(a) p–type silicon wafer (substrate).

(b) n–type silicon epitaxial layer grown on to p-substrate.

(c) p–type acceptor impurity diffused in, to start isolation region.

(d) p+ impurity diffused in, to form top gate. Simultaneously, p region diffuses to complete isolation between the n islands.

(e) n++ type impurity diffused into n-type channel to form Drain and Source contacts. Physical isolation of each FET is achieved by scribing and breaking wafer along X–X axis.

(f) Formation of discrete n-channel FETS after 'scribing and breaking' procedure. Since the edges of the p+ regions of the top gate are diffused into the p-substrate, electrical contact is made to the top gate, via the p-substrate.

FIGURE 1-4   Some assembly steps for *n*-channel JFETs.

1-3-2 The Small-Signal Planar MOSFET

Figure 1-5 shows some of the steps involved in the manufacture of one type of planar, small-signal enhancement-mode MOSFET. The starting material, as with the JFET, is a wafer of high-purity, p-type, monocrystalline silicon, 1-5a, onto which a layer of silicon dioxide has been deposited, 1-5b. Prior to about 1980 discrete small-signal MOSFETs were manufactured with metal gates. This process technology has been largely replaced with a polysilicon process that offers vastly improved reliability, and easier fabrication. To achieve a polysilicon gate an n-doped polysilicon layer is grown over the oxide, 1-5c. By suitable masking and photolithographic technique the gate is formed, 1-5d. Using the poly gate as a mask, an ion-implant penetrates the silicon dioxide placing deep n-doped regions straddling the gate, 1-5e. These implants become the source and drain of the MOSFET once metallic contacts have been attached, 1-5f.

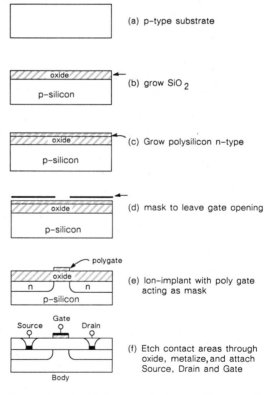

FIGURE 1-5 Fabrication of planar enhancement-mode MOSFET.

1-3-3 The DMOSFET (Lateral and Vertical)

Two significant techniques are being utilized to produce short-channel MOSFETs. Both utilize the difference in depth of two diffusions instead of photographic techniques to control channel length. The differences in the two techniques relate to the geometry employed. In one type (the lateral structure) the channel is oriented parallel to the chip surface (as with the previously described MOSFET). The second type, popularized by the burgeoning power MOSFET market, has its channel oriented as an inverted L. This type has the source on the surface and the drain beneath the chip. Despite the apparent differences in manufacture between these two types both are DMOS (for diffused-channel MOS).

Manufacture of the lateral structure, shown in Fig. 1-6, begins with a p-doped monocrystalline wafer 1-6a, onto which is first grown a layer of silicon dioxide 1-6b, followed by a layer of silicon nitride 1-6c. A mask, 1-6d, defines the gate region when the silicon nitride is etched away, 1-6e, leaving the gate. Using the silicon nitride gate and additional masking the p-doped channel is formed by ion implant, 1-6f. Removing the mask

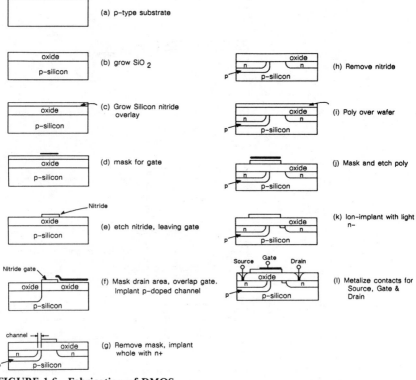

FIGURE 1-6   Fabrication of DMOS.

allows an n^+ ion implant to form both the source and drain 1-6*g*. The polysilicon is then grown over the oxide 1-6*h*, and suitably masked 1-6*i*, and etched leaving a polysilicon gate that extends well over the n-doped source toward the drain implant 1-6*j*. Using this poly as a mask an ion implant of light n^- places a thin channel beneath the oxide 1-6*k*. All that remains is to metalize the n-implanted source, drain and poly gate to accept the leads 1-6*l*.

The construction of the power MOSFET differs dramatically from either the small-signal planar MOS or the lateral DMOS, as we see in Fig. 1-7. Where, for an n-type DMOS we began with p-doped semiconductor,

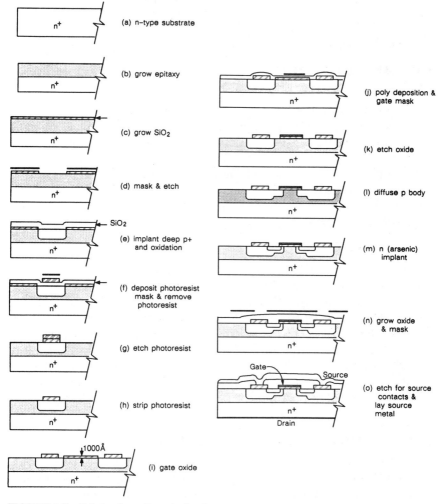

FIGURE 1-7 Fabrication of vertical enhancement-mode power MOSFET.

for the n-type power MOSFET we start with heavily-doped n-doped semiconductor 1-7a into which we grow a thick lightly doped n^- epi layer 1-7b. The thickness of this n^- epi plays an important role in establishing the ultimate breakdown voltage (and ON resistance) of the power MOS-FET. The steps shown in Fig. 1-7 are well-documented and with careful study should be reasonably clear even to those unfamiliar with semiconductor processing. However, for clarity, we need to note Fig. 1-7o where metal appears to cover the surface of the power MOSFET seemingly isolating the gate. We must take note that a surface view would show the source metal in a symmetrical pattern with openings for contact to the polysilicon that runs checkerboard fashion over the surface.

1-4   HOW SMALL-SIGNAL FETS WORK

1-4-1   The Junction FET

n-Channel JFET

A section view of a JFET is shown in Fig. 1-8. This structure contains, between the source and the drain contacts, an n-type "channel" embedded in a p-type silicon substrate. If it is assumed that the p-n junction forms a barrier to current flow, then it can be seen that channel conduction is a function of the channel width, length, and thickness and of the density and mobility of the carriers. In this structure current can flow equally well in either direction through the channel; i.e., the drain may be positive or negative with respect to the source.

p-Channel JFET

The foregoing discussion of the JFET considered the n-type channel. A p-channel FET has similar characteristics, with the major difference being a change in the polarities. Characteristic curves of a p-channel device are given in Fig. 1-9. P-type carriers (holes) have lower mobility than n-type carriers (electrons); thus for devices of similar dimensions and channel carrier density, the n-type FET will have a higher channel conductance than the p-type device. For a given channel conductance and pinchoff voltage, the p-channel FET will typically have higher interelectrode capacitance and higher junction leakage than the n-channel device.

1-4-2   The MOSFET (Insulated-Gate FET)

A cross section of an n-channel depletion-mode MOSFET is shown in Fig. 1-10. With this structure the n-channel conduction may be controlled by a voltage applied between gate and source, by one applied between body and source, or by a combination. A very important characteristic of

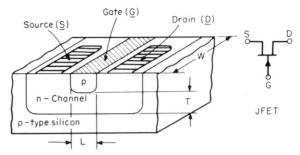

FIGURE 1-8   Junction-type field-effect transistors.

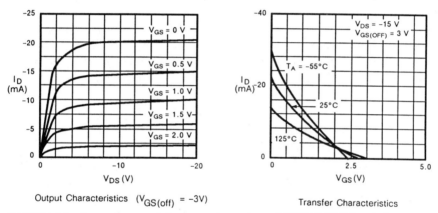

Output Characteristics ($V_{GS(off)}$ = –3V)

Transfer Characteristics

FIGURE 1-9   Some characteristic curves for p-channel JFET (2N5114-series).

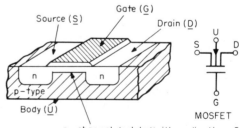

n – channel depleted with application of negative gate voltage. Negative body voltage also depletes the channel.

FIGURE 1-10   Cross-section of n-channel MOS-FET.

this structure is that the gate is separated from the channel by a very low-leakage dielectric (typically silicon oxide). This permits the gate-channel voltage to be positive or negative. For the n-channel structure of Fig. 1-10, a positive gate-source voltage will increase channel conduction, while a negative voltage will decrease channel conduction, as shown in Fig. 1-11. Since no conducting path exists between the control gate and the rest of the structure, the gate-to-channel resistance of a typical device exceeds 10^{16} Ω. This is true with positive or negative voltage applied to the gate. An upper limit on gate voltage is imposed by the dielectric breakdown of the thin oxide under the gate metal. Unlike in a p-n junction, the breakdown of the oxide dielectric results in permanent damage to the oxide; thus this condition must be avoided.

A normally OFF (enhancement-mode) MOSFET is shown in Fig. 1-12. In this structure no channel exists between drain and source unless a gate voltage is applied. A positive voltage applied to the gate, with respect to source and drain, creates a conducting channel by pulling carriers (electrons for the n-channel type, holes for the p-channel type) from the source and drain regions and from the body into the upper layers of the substrate under the gate. The thickness and conduction of the channel

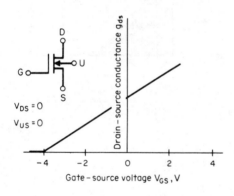

FIGURE 1-11 Channel conductance vs. gate voltage for n-channel MOSFET.

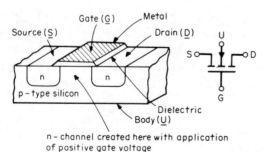

n – channel created here with application of positive gate voltage

FIGURE 1-12 An n-channel enhancement-mode MOSFET— normally OFF.

thus created are a function of the gate-source, gate-drain, and gate-body voltages. The gate-source voltage V_{GS} at which channel conduction just begins to occur is called the "threshold voltage" $V_{GS(th)}$. $V_{GS(th)}$ is a function of the thickness of the oxide under the gate, p-type carrier density in the body under the gate, and body-source voltage V_{US}. The body could be designed to function as a junction-type control gate but is usually designed to have minimum effect upon the channel conduction and threshold voltage. In many device types the body is internally connected (shorted) to the source terminal, in which case the source-drain voltage is limited to one polarity to avoid forward-biasing the drain-body diode. Figure 1-13 shows channel conduction g_{ds} versus V_{GS} with V_{US} as a parameter. With this particular device, V_{US} has a great effect upon $V_{GS(th)}$ and g_{ds}. In many device designs the body effect is less than that shown.

Dual-Gate MOSFET

A useful structure for high-frequency applications is the dual-gate MOSFET. This device, as shown in Fig. 1-14, has two control gates between

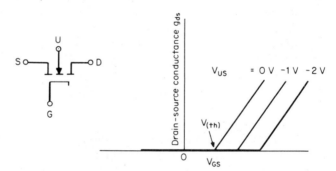

FIGURE 1-13   Channel conductance vs. gate-source voltage for a MOSFET showing the effect of body bias.

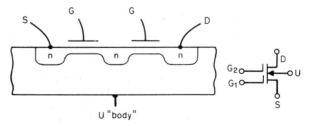

FIGURE 1-14   Dual-gate MOSFET.

the drain and the source. Typically in a high-frequency amplifier, G_1 is utilized as the signal input terminal and G_2 is at signal ground. This results in very low drain-to-G_1 feedback capacitance. This operating mode is similar to the "Cascode" operation of vacuum tubes. G_2 is also commonly utilized as an automatic gain control element.

1.5 SUMMARY

This chapter has given a qualitative description of several types of field-effect transistors. As the presentation is aimed at the FET user, device design details have been avoided.

Chapter 2 presents a more detailed electrical description of the FET, including mathematical interrelationships of the various characteristics such as pinchoff voltage, drain saturation current, transconductance, channel ON resistance, temperature coefficients, and inner-electrode capacitance.

Other chapters will be devoted to applications of FETs in such circuit functions as analog switches, amplifiers, mixers, oscillators, voltage converters, voltage-controlled resistors, and constant current sources.

Although the commercial availability of FETs followed the bipolar transistor by 8 to 10 years, this device has become a very valuable component in electronic circuits. Integrated circuits in such devices as computers, memories, and electronic TV games make extensive use of MOS-FETs. TV and FM tuners and hi-fi amplifiers utilize FETs as input stages because of their superior low-noise performance. Ionization-chamber-type smoke detectors utilize a MOSFET input-stage amplifier because of the very low leakage of the MOSFET gate. Analog multiplexer systems incorporate FETs and analog switches because of the absence of junction offsets in the ON state and the bidirectional blocking capability in the OFF state. Recent developments in the manufacture of short-channel FETs have increased power capabilities and have resulted in increasing high-power, high-frequency applications, a few of which are presented in the chapters that follow.

BIBLIOGRAPHY

Hauser, J. R.: "Characteristics of JFET Devices with Small Channel Length-to-Width Ratios," *Solid State Electronics*, **10**:577–587, 1967.

Oxner, Edwin S.: *FET Technology and Applications, An Introduction*. Marcel Dekker, Inc., New York, 1989.

Shockley, W.: "A Unipolar Field Effect Transistor," *Proceedings, IRE*, **40**:1365–1376, 1952.

Sze, S. M.: *Physics of Semiconductor Devices*, John Wiley and Sons, Inc., New York, 1969.

2

PARAMETERS AND SPECIFICATIONS

2-1 INTRODUCTION

Chapter 1 was intended primarily to provide an introduction to the various types of FETs rather than to serve as an analysis of their behavior. In this chapter we examine their parameters, both static and dynamic, from which the distinctive specifications are derived. From a practical standpoint, it is often preferable to represent the device characteristics by experimentally determined curves rather than by exact mathematical expressions, because the effects of donor and acceptor impurity concentrations, initial channel and depletion-layer thickness, photographic mask configurations and alignments, and variations of these within a device type are thus taken into account. The equations presented in this chapter should not be considered exact, but do give close approximations of the performance of most devices. They follow, for the most part, FET theory presented by Shockley[1] and others; however, certain physical characteristics assumed in theory are sometimes difficult to achieve in practice. For example, the original Shockley theory assumed abrupt junctions, uniform channel conductivity, and a carrier mobility independent of electric field intensity. Real devices, made by diffusion techniques, have some gradation of junctions and channel conductivity. In short-channel FETs the field at the drain end of the channel is typically high enough to cause some reduction in carrier mobility.

For the initial discussions about static characteristics and the interrelationship of various parameters, we will use as an example a device which has a relatively long channel. Later the effect of shortening the channel length will be discussed.

[1]Shockley, W., "A Unipolar Field-Effect Transistor," *Proceedings, IRE* **40**:1365–1376, 1952.

2-2 THE SMALL-SIGNAL JFET

In general, the junction FET is characterized by both its static and dynamic parameters. *Static* being defined as dc parameters, such as the breakdown and pinchoff voltages, the saturation drain current and ON resistance (or channel conductance), and the operating gate current. The *dynamic* characteristics are defined by the ac parameters. These include transconductance, the parasitic interelectrode capacitances, switching characteristics, and high-frequency parameters as applicable and most of the operating parameters including, for example, the noise parameters.

2-2-1 Basic Static Parameters

The five basic static parameters are breakdown voltage $V_{(BR)GSS}$, gate cutoff voltage $V_{GS(off)}$, saturation drain current I_{DSS}, channel conductance (ON resistance), and operating gate current I_G.

Shockley's power-law equation, although more appropriate for explaining the dynamic characteristics, nevertheless shows the relationship between bias and drain current for the depletion-mode (junction) FET.

$$I_D = I_{DSS}\left(1 - \frac{V_{GS}}{V_{GS(off)}}\right)^n \tag{2-1}$$

This equation indicates that for $V_{GS} = V_{GS(off)}$, $I_D = 0$. In practice this does not happen. Starting from zero, as the gate voltage is made more negative, the drain current decreases until it reaches a very low value equal to the drain-gate leakage current. At this value the source current will consist of source-gate leakage. Any further increase in the magnitude of the negative gate voltage will result in an increase in I_D leakage.

We can perhaps visualize the phenomenon of the I_D-versus-V_{GS} characteristic in Fig. 2-1 in which the data are plotted logarithmically to show the current magnitude near "cutoff." Because I_D does not go to zero, the error of Eq. (2-1) increases as V_{GS} approaches $V_{GS(off)}$. From a practical measurement standpoint, $V_{GS(off)}$ is usually specified at an I_D value greater than the minimum expected leakage current. The symbol $I_{D(off)}$ is used for the approximate minimum value of I_D. Device types characterized for switching applications will usually specify both $V_{GS(off)}$ and $I_{D(off)}$, as shown in the example for the 2N3970 specification:

Characteristic	Min	Max	Unit	Test conditions
$I_{D(off)}$ drain cutoff current		250	pA	$V_{DS} = 20$ V, $V_{GS} = 12$ V
$V_{GS(off)}$ gate-source cutoff voltage	−4	−10	V	$V_{DS} = 20$ V, $I_D = 1$ nA

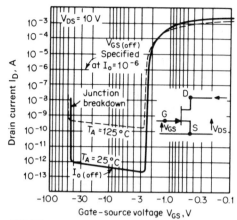

FIGURE 2-1 Drain current vs. gate-source voltage for *n*-channel JFET (2N4868A).

For this device $V_{GS(\text{off})}$ is specified as that value of V_{GS} required to reduce I_D to 1 nA, with a V_{DS} of 20 V.

Figure 2-1 shows that at a very high negative value of V_{GS}, there occurs a rapid increase in I_D. This is the voltage at which avalanche "breakdown" occurs at the drain-gate n-p junction; it sets an absolute maximum rating on the drain-gate voltage for the device. In practice, the device data sheet will usually specify a minimum value for gate-to-drain and gate-to-source breakdown, as shown in the 2N3970 specification:

Characteristic	Min	Max	Unit	Test conditions
$V_{(\text{BR})GSS}$ gate reverse breakdown voltage	−40		V	$I_G = 1\ \mu\text{A},\ V_{DS} = 0$

Since the test conditions set $V_{DS} = 0$, this specification ensures that both drain-to-gate and source-to-gate breakdown will be equal to or greater than 40 V. The value of I_G for the test condition is greater than the normal gate leakage current of the device. I_G-versus-V_{GS} curves are shown in Fig. 2-2 with test points for $V_{(\text{BR})GSS}$ and I_{GSS} indicated. I_{GSS} is the gate leakage current at a specified value of V_{GS} with $V_{DS} = 0$. Forward gate characteristics are not usually specified because few applications require operation in this mode. The curve of Fig. 2-2*b* is typical for a small-signal JFET with the gate *forward*-biased. Operation with a few tenths of a volt

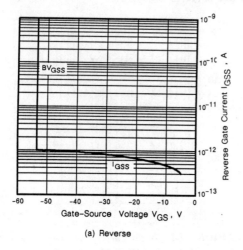

(a) Reverse

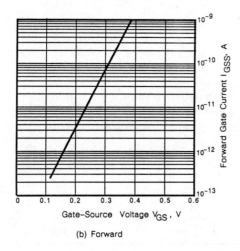

(b) Forward

FIGURE 2-2   Gate current vs. gate-source voltage.

forward gate signal is satisfactory if a few hundred picoamperes of gate current permits proper circuit performance.

Breakdown Voltage

The maximum voltage is limited by the gate-to-channel breakdown voltage. In JFETs breakdown results from avalanche multiplication of carriers in the depletion region occurring between the gate-to-drain end of the channel. As a consequence the drain-gate breakdown depends on the gate-source voltage as seen in Fig. 2-3.

Breakdown is strongly dependent upon the shape of the JFET junctions, in particular at the corners of curvature between the p-gate and the

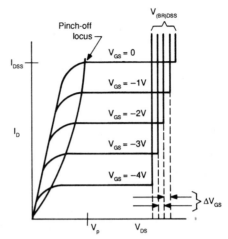

FIGURE 2-3 Effect of gate-source voltage on breakdown.

n-channel at the drain end. Studies[2] show that the breakdown voltage decreases significantly as the junction depth (curvature) increases.

Most JFETs in use today exhibit breakdown voltages ranging from 20 to 80 V.

Channel Conductance (g_{ds})

For any FET the channel conduction is controlled by a gate-to-channel voltage. As indicated in Fig. 2-4a, at the p-n (gate-to-channel) junction there is a depletion layer. If there is an abrupt change from a high concentration of holes in the gate region (p-type) to a much lower concentration of electrons in the channel region (n-type), then most of the depletion width will occur in the channel. The depletion width is proportional to the square-root of the junction potential.[3]

$$W_d = \left(\frac{V_{bi} \pm V_{GS}}{K_1 N_c} \right)^{1/2} \tag{2-2}$$

where W_d = junction depletion width
  V_{bi} = built-in junction potential
  V_{GS} = reverse or forward bias applied to gate-channel junction
  K_1 = constant
  N_c = channel carrier concentration.

[2]Leistiko, O., and A. S. Grove, "Breakdown Voltage of Planar Silicon Junctions," *Solid-State Electronics* **8**:847–852, 1966.

[3]Shockley, W., *op. cit.*

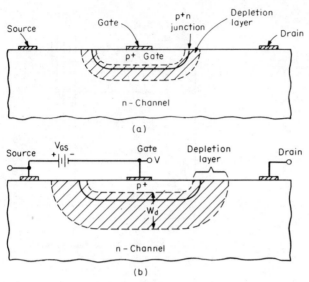

FIGURE 2-4   Cross-sectional view of JFET showing the depletion layer at the p^+ n junction.

The channel conduction is decreased (ON resistance increased) as the junction depletion width is increased and thus can be controlled by the gate-source voltage V_{GS}.

The channel conduction g_{ds} of a typical JFET as a function of the junction voltage is shown in Fig. 2-5. The characteristic is shown with zero volts between the "source" and the "drain;" thus the depletion thickness is uniform along the channel length. The voltage at which the channel conduction approaches zero is the "gate cutoff voltage" and is given by the symbol $V_{GS(off)}$. The polarity of $V_{GS(off)}$ is such that the p-n gate-channel junction is *reverse*-biased. A forward bias at the gate junction will increase channel conduction (lower ON resistance) because of the resulting decrease in depletion layers; however, as indicated in Fig. 2-6, a *forward* gate-channel bias results in a large increase in gate current. In the FET symbol shown in Fig. 2-5, note that the direction of the arrow at the gate-channel junction shows the direction of easy gate-current flow. (This is similar to the p-n diode symbol.) In its reverse-bias mode, the gate "leakage" current may be only a few picoamperes; thus the low-frequency input impedance of the FET is very high. In most JFET applications the forward gate-bias mode is not used because of the resulting low input impedance. At some relatively high reverse gate-channel voltage, avalanche breakdown will occur at the p-n junction. This places an upper limit on the device operating voltage.

Figure 2-7 presents a concept of the depletion in the channel at various

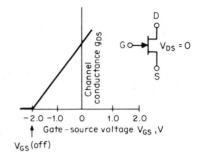

FIGURE 2-5 **Channel conductance vs. gate-channel voltage.**

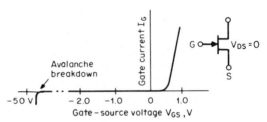

FIGURE 2-6 **Gate current vs. gate-source voltage.**

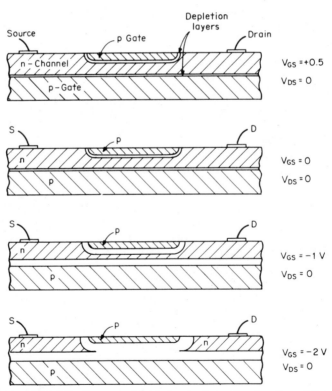

FIGURE 2-7 **Concept of depletion in channel as function of the gate-to-channel voltage** $(V_{DS} = 0)$.

gate-channel voltages. These sketches assume zero drain current so that there is no voltage drop along the channel length. If a positive voltage is applied to the drain, with the gate-source voltage at zero, a nonuniform depletion width occurs along the channel as shown in Fig. 2-8. As V_{DS} is made more positive, the depletion layer at the drain end of the channel increases, and thus the incremental channel conduction decreases; if the channel is long (compared to its thickness), then the channel will "pinch-off" at the drain end and channel current will saturate at a drain-gate voltage approximately equal to $-V_{GS(off)}$. Further increases in drain voltage will have little effect upon drain current, hence the current "saturates." This effect is known as the saturated drain current I_{DSS}.

Figure 2-9 shows the drain current I_D versus drain-source voltage V_{DS}, with the gate-source voltage V_{GS} at zero. As V_{DS} is made more positive, the depletion layer at the drain end of the channel increases, and thus the incremental channel conductance decreases. The slope dI_D/dV_{DS} continues to decrease as V_{DS} is increased, until at some high voltage breakdown occurs. In the negative quadrant a rapid increase in drain current occurs because the drain-gate junction becomes forward-biased. The JFET is not normally operated in this quadrant more than a few tenths of a volt. Figure 2-10 shows the magnitude of dI_D/dV_{DS} as a function of V_{DS} for a

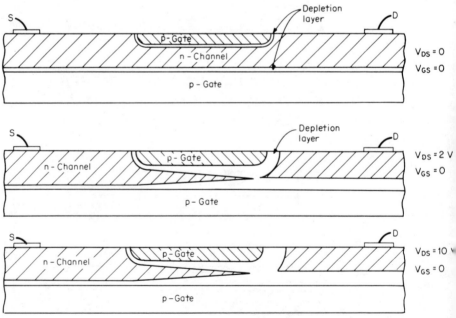

FIGURE 2-8   Channel depletion vs. drain-source voltage ($V_{GS} = 0$). Compare with Fig. 2-7.

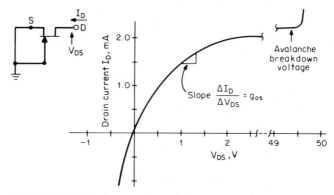

FIGURE 2-9   Drain current vs. drain-source voltage.

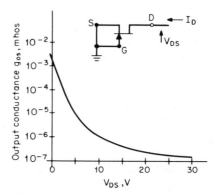

FIGURE 2-10 Output conductance vs. drain-source voltage.

typical FET. This is the output conductance g_{os} of the FET operating in a common-source configuration.

With reference to Fig. 2-8, it may appear that when V_{DS} exceeds "pinch-off," no current can flow. However, there does exist an electric field across this "depletion" region and a supply of carriers (electrons) at the source end of the channel; thus carriers will drift across the depletion region just as they drift across the collector-base depletion region in a bipolar transistor. As V_{DS} approaches the pinchoff value the drain current I_D tends to limit at a saturation level.

Effect of Channel Length

The degree of saturation is a function of the device geometry. A long channel will become more saturated (have a lower dI_D/dV_{DS}) than a short channel; however, the long channel will also have a lower initial channel

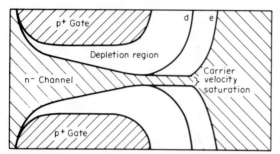

FIGURE 2-11 Shape of conductive channel when carrier velocity saturation occurs. (Reprinted with permission from *Solid State Electronics*, vol. 10, J.R. Hauser, "Characteristics of JFET Devices with Small Channel Length-to-Width Rations," © 1967 Pergamon Press, Ltd.)

conductance than will the short channel. For the short-channel device (where length-to-thickness ratio is less than about 4), drain-voltage "pinchoff" may be higher than $-V_{GS(off)}$ ($|V_p| > |V_{GS(off)}|$). This is because the minimum channel width does not occur directly under the gate junction but is shifted toward the drain terminal. A concept of the shape of a short channel versus drain voltage is shown in Fig. 2-11. This figure shows the channel shape when carrier velocity saturation occurs. Most devices designed for high-frequency and high-power applications have short channels. For these devices the drain-current saturation I_{DSS} may be due to carrier velocity saturation instead of channel "pinchoff." It has been calculated[4] that for a channel length-to-thickness ratio of unity, drain voltage V_p at which saturation occurs will be approximately 1.6 times $-V_{GS(off)}$.

For the short-channel FET an effect known as carrier "velocity saturation" may cause drain current to saturate before minimum channel width is reached. This effect occurs because at high electric fields, carrier mobility is no longer constant but becomes inversely proportional to the electric field. For fields greater than this critical value, drift velocity no longer increases with increasing field; hence the drain current saturates.

Gate-Source Cutoff Voltage ($V_{GS(off)}$)

Figure 2-7 presents a concept of the depletion in the channel at various gate-channel voltages. The depletion region is controlled by the gate-

[4]Hauser, J. R., "Characteristics of JFET Devices with Small Channel Length-to-Width Ratios," *Solid-State Electronics* **10**:577–587, 1967.

source voltage V_{GS}. The depletion region widens as V_{GS} becomes more negative decreasing the channel conductance. Therefore, for values of V_{DS} near zero volts, the channel conductance is controlled solely by V_{GS}. As V_{GS} increases (more negatively for an n-channel JFET) a critical voltage is reached where no drain current flows. This critical voltage is the cutoff voltage, $V_{GS(off)}$.

The magnitude of this cutoff voltage, $V_{GS(off)}$, is, among other things, a function of the gate diffusion depth. The deeper the diffusion the less bias is required to achieve cutoff. This is easily visualized in Fig. 2-12.

The gate-source cutoff voltage is an important parameter since many of the other JFET parameters may be predicted from it. Derivations of Shockley's equation (Eq. 2-1) reveal that the saturation drain current I_{DSS} and channel conductance may be derived from a knowledge of $V_{GS(off)}$. A plot showing the relationship is provided in Fig. 2-13.

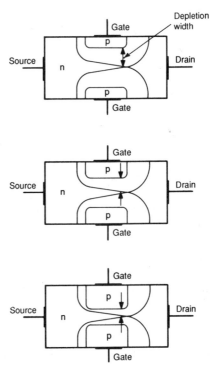

FIGURE 2-12 The deeper the gate diffusion the less gate voltage is required to achieve cutoff.

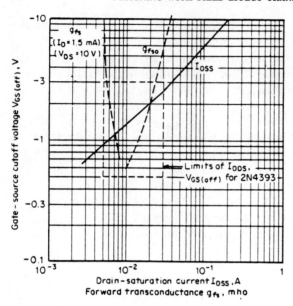

FIGURE 2-13 The relationship between I_{DSS} and g_{fs} vs. $V_{GS(off)}$ for 2N4393 JFET.

Gate-Current Characteristics

The gate current I_G of a JFET not only is a function of V_{GS} and V_{GD} but also may be a function of I_D. Figure 2-14d shows gate-current characteristics of type 2N4868A. This device has a $V_{(BR)GSS}$ in excess of 40 V; however, when the device is biased such that drain-to-source current is permitted to flow (a normal amplifier condition), then the gate-current "breakpoint" occurs at a lower drain-to-gate voltage. This breakpoint is a function of basic device design and is usually in the 8-V to 30-V range. Beyond this breakpoint, I_G is approximately a *linear* function of I_D and an exponential function of V_{DS}.

An understanding of the I_G-versus-V_{DG} characteristics is aided by considering the equivalent circuit shown in Fig. 2-15, which indicates that I_G is the sum of three components: I_1, I_2, and I_3. I_1 and I_2 are simple leakage currents of reverse-biased gate-source and gate-drain p-n junctions. They are the result of *thermal* ionization of carriers within the junction depletion layers and will approximately double in magnitude for each 10°C increase in junction temperature. For both n-channel and p-channel JFETs the value of I_1 and I_2 below the breakpoint is approximately proportional to $(V_{GS})^{1/2}$ and $(V_{GD})^{1/2}$.

The I_3 component results from carriers generated within the drain-

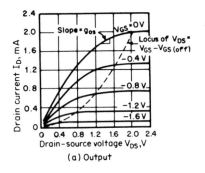

(a) Output

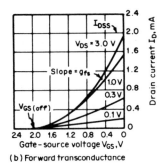

(b) Forward transconductance

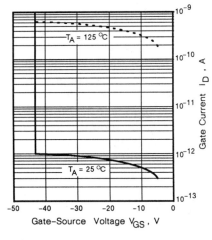

(c) Input

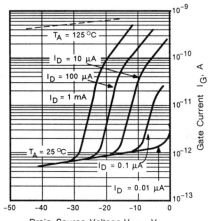

(d) Reverse transconductance

FIGURE 2-14 Various characteristics of n-channel JFET.

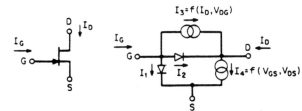

(a) FET symbol   (b) Equivalent circuit

FIGURE 2-15   Equivalent circuit of JFET useful to analyze gate current.

gate space-charge region due to *impact* ionization by the drain-current carriers. Thus I_3 (the I_G mentioned earlier) is a linear function of I_D and an exponential function of V_{DG}. At low voltage the impact ionization rate is so low that the I_3 component is undetectable in the presence of the thermally generated components I_1 and I_2. Impact ionization rate, however, is an exponential function of the electric field. Thus increasing V_{DG} causes a rapid increase in the I_3 component.

Impact ionization is also a function of carrier mobility. Since mobility has a negative temperature coefficient, I_3 *decreases* with increasing temperature. The broken-line in Fig. 2-14c shows I_G versus V_{DG} at a temperature of 125°C. Note that the breakpoint appears to occur at a higher value of V_{DG} than at 25°C.

This I_G breakpoint dependency upon I_D occurs at higher voltages in p-channel JFETs because of the lower mobility and lower ionization rates of holes in the drain space-charge region.

In most applications where a low I_G is important, the FET will be biased below the "knee" or "breakpoint" of the I_G-versus-V_{DG} curve.

Gate operating current I_G should *not* be considered equal to the reverse-biased gate-junction leakage current I_{GSS} that is characterized at $V_{DS} = 0$. The curves in Fig. 2-16 show how I_G breakpoint is related to basic device design. JFETs with a high g_{fs}/C_{iss} ratio typically have low breakpoint voltages, whereas high-μ devices ($\mu = r_{ds} \times g_{fs}$) have much higher I_G breakpoints, typically $V_{DG} = 20$ to 30 V.

To minimize I_G under operating conditions, particular attention must be paid to V_{DG}. The critical drain-gate voltage (I_G breakpoint voltage) can

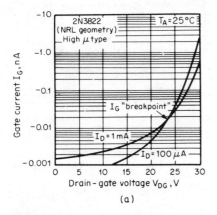

(a)

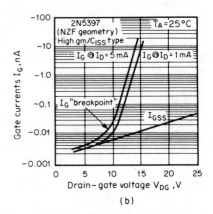

(b)

FIGURE 2-16 Operating gate current as function of drain-gate voltage and drain current. These curves, in essence, represent the reverse breakdown characteristics (3rd quadrant) of the common *pn* diode.

be anywhere from 8 to 30 V. Should a common-mode voltage range in excess of 10 to 15 V be required, a cascode input stage should be considered to ensure that the first JFET has a low drain-gate voltage. In most applications where a low I_G is important (for *all* high input resistance and low noise applications), the JFET will be biased below the "knee" of "breakpoint" of the I_G-versus-V_{DS} curve. If this is the case, then I_G will approximately double for every 10 to 12°C increase in temperature.

2-2-2 Basic Dynamic Characteristics

Because FETs are employed largely in circuits having varying currents and voltages, their dynamic characteristics are of interest.

Shockley's equation (Eq. 2-1) and its derivatives provides a comprehensive understanding of many of the *basic* dynamic characteristics of the JFET with the exception of the parasitic capacitances.

Forward Transconductance

A useful circuit design characteristic is the effect of gate-source voltage upon drain current. A family of these I_D versus V_{GS} "transconductance" characteristics for a long-channel FET is shown in Fig. 2-17. The data for these curves were obtained using a 2N4868 that had a $V_{GS(off)}$ of approximately -2 V. If $|V_{DS}| > |-V_{GS(off)}|$ by differentiating Eq. (2-1), the small-signal transconductance g_{fs} is given by

$$g_{fs} = \frac{dI_D}{dV_{GS}} = \frac{I_{DSS}}{V_{GS(off)}} \left(1 - \frac{V_{GS}}{V_{GS(off)}} \right)^{n-1} \tag{2-3}$$

Some texts indicate a value of 3/2 for n; however, experimental measurements on a number of n-channel JFET geometries indicate that the

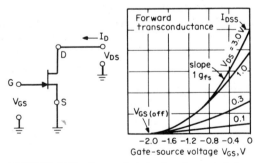

FIGURE 2-17 Forward transfer characteristics of n-channel JFET.

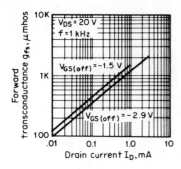

FIGURE 2-18 Common-source forward transconductance vs. drain current.

exponent n is closer to 2, which is the value derived in a treatment by R. D. Middlebrook[5] from the ratio of Eqs. (2-3) and (2-1):

$$\frac{g_{fs}}{I_D} = n \left(V_{GS} - V_{GS(\text{off})} \right)^{-1} \qquad (2\text{-}4)$$

At $V_{GS} = 0$, $I_D = I_{DSS}$ and $g_{fs} = g_{fso}$, using 2 as the value of the constant n leads to,

$$g_{fso} = -2 \, \frac{I_{DSS}}{V_{GS(\text{off})}} \qquad (2\text{-}5)$$

For n-channel FETs I_{DSS} is positive and $V_{GS(\text{off})}$ is negative; for p-channel FETs I_{DSS} is negative and $V_{GS(\text{off})}$ is positive; thus g_{fs} is a positive quantity for both p- and n-channel FETs.

Referring to Fig. 2-14b, the slope of the curve or rate of change of I_D with respect to V_{GS} is important. This slope is the forward (gate-to-drain) transconductance of the FET and is given the symbol g_{fs}. As is apparent in Fig. 2-14b, g_{fs} is a function of I_D, decreasing as I_D is decreased. Figure 2-18 shows the relationship between g_{fs} and I_D for a typical FET type. Curves for two devices of the same geometry but having different $V_{GS(\text{off})}$ values are included to show that g_{fs} is also a function of $V_{GS(\text{off})}$ as Eqs. (2-4) and (2-5). When g_{fso} does not equal g_{fs} (when $V_{GS} \neq 0$) then,

$$g_{fs} = g_{fso} \left(\frac{I_D}{I_{DSS}} \right)^{1/2} \qquad (2\text{-}6)$$

The exponent in Eq. (2-6) is indicated by the slope of the curves of Fig. 2-18.

These curves indicate that g_{fs} is an approximate function of $(I_D)^{1/2}$. This square-root relationship is common for long-channel JFETs and means

[5]Middlebrook, R. D., "A Simple Derivation of Field-Effect Transistor Characteristics," *Proceedings, IEEE* **50**:1146–1147, 1963.

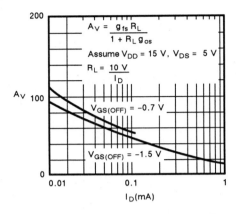

FIGURE 2-19 Voltage gain is increased by lowering drain current if the product $I_D R_L$ remains fixed. (Courtesy of Siliconix Incorporated.)

that in an RC-coupled amplifier, voltage gain may be increased by operating at lower drain currents, if $I_D R_L$ is kept constant (a "starved" amplifier), as shown in Fig. 2-19.

Output Conductance

The output characteristic curve for $V_{GS} = 0$ of Fig. 2-14a approximately follows the equation

$$I_D = g_{dso} V_{DS} \left(1 + \frac{V_{DS}}{2 V_{GS(off)}}\right) \tag{2-7}$$

where $g_{dso} = $ slope dI_D/dV_{DS} at V_{GS} and $V_{DS} = 0$. However, g_{fso}; therefore, by utilizing Eq. (2-5), Eq. (2-7) can be given as

$$I_D = \frac{2 I_{DSS}}{-V_{GS(off)}} V_{DS} \left(1 + \frac{V_{DS}}{2 V_{GS(off)}}\right) \tag{2-8}$$

If we rearrange terms and for convenience let $V_p = V_{GS(off)}$, we get

$$I_D = I_{DSS} \left[2 \frac{V_{DS}}{-V_p} - (\frac{V_{DS}}{V_p})\right]^2 \tag{2-9}$$

This equation is not valid when V_{DS} exceeds $-V_p$. It indicates that as V_{DS} approaches $-V_p$, the rate of change of I_D with changing V_{DS} decreases and reaches zero at $V_{DS} = V_p$. In a real device dI_D/dV_{DS} does not reach zero.

For some types of applications of the FET, it is helpful to understand the characteristics at very low values of V_{DS}, such as those shown in Fig.

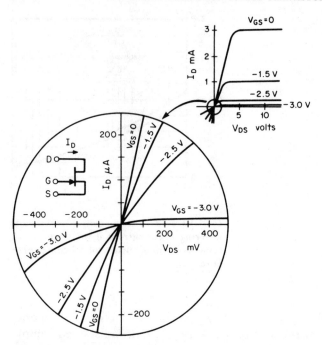

FIGURE 2-20 n-channel JFET output characteristic enlarged around $V_{DS} = 0$.

2-20. A "very low value" is one that is small compared to the magnitude of $V_{GS} - V_{GS(off)}$ (for an n-channel JFET).[6] In this region V_{DS} is small enough to have little effect upon channel thickness, so that the I_D/V_{DS} slope is nearly linear. Since the slope is a function of V_{GS}, the FET can be utilized as a voltage-controlled resistor (VCR). The conductance slope (dI_D/V_{DS}) at $V_{DS} = 0$ is approximately a linear function of $V_{GS} - V_{GS(off)}$.

If g_{dso} at $V_{GS} = 0$ is given, then

$$g_{ds} = g_{dso}\left(1 - \frac{V_{GS}}{V_{GS(off)}}\right) \tag{2-10}$$

with $V_{DS} = 0$. A plot of this characteristic is shown in Fig. 2-21 along with g_{fs} and I_D characteristics.

The relationship between g_{dso}, $V_{GS(off)}$ and I_{DSS} at very low values of V_{DS} ($V_{DS} \ll V_p$) is given by Eq. (2-5) (under which conditions $g_{dso} = g_{fs}$), where I_{DSS} and $V_{GS(off)}$ are as indicated in Fig. 2-14b. It is important to note that

---

[6]For a p-channel JFET the "low value" must be compared to the magnitude of $V_{GS(off)} - V_{GS}$.

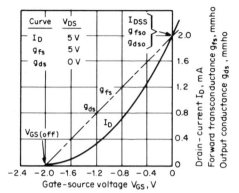

FIGURE 2-21 I_D, g_{fs}, and g_{ds} vs. gate-source voltage.

Eq. (2-5) and (2-10) are valid only for the case where V_{DS} is very small compared to V_p.

Conductance Effects on Gain

At $V_{DS} = 0$, g_{os} is equal to the g_{dso} discussed in association with Eq. (2-10). As V_{DS} is increased, g_{os} decreases. For voltage-amplifier applications, a low g_{os} is of importance because the maximum voltage gain is limited by the ratio g_{fs}/g_{os}. If I_D really saturated at $-V_{GS(off)}$, then g_{os} would drop to zero at V_{DG} equal to or greater than $-V_{GS(off)}$. Most device types designed for small-signal low-frequency amplifiers will have g_{fs}/g_{os} ratios in excess of 100 (for $V_{DS} > -V_{GS(off)}$). Figure 2-22 shows g_{os} versus V_{DS} for typical n-channel JFETs. Figure 2-23 shows g_{os} versus I_D for a constant value of V_{DS}. The curves of Figs. 2-18, 2-22, and 2-23 reveal two important points about devices of a given basic design. First, when operated at a given value of I_D, the device with the lower $V_{GS(off)}$ will have a higher g_{fs} and a

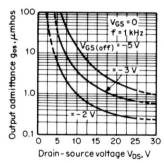

FIGURE 2-22 Output admittance vs. drain-source voltage.

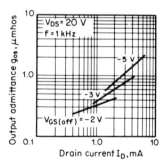

FIGURE 2-23 Output admittance vs. drain current.

lower g_{os}. Second, for a given power-supply voltage, available voltage gain is inversely proportional to $V_{GS(off)}$. Also, lower values of $V_{GS(off)}$ permit use of lower-voltage supplies.

A simplified common-source FET voltage-amplifier circuit and its low-frequency equivalent circuit are shown in Fig. 2-24. This equivalent circuit neglects reactive components and dc leakages. It is given here to show the equivalent circuit location of the g_{fs} and g_{os} parameters discussed above. The voltage amplification with this circuit is

$$A_v = \frac{e_o}{e_g} - g_{fs}\left(\frac{R_L}{1 + g_{os}R_L}\right) \tag{2-11}$$

If g_{os} is small compared to $1/R_L$, then Eq. (2-11) reduces to

$$A_v = -g_{fs}R_L \tag{2-12}$$

The negative sign indicates that the phase of the signal is inverted.

The low-frequency input conductance g_{iss} contains two principal components: the gate-source conductance ($g_{gs} = dI_G/dV_{GS}$) and the gate-drain conductance ($g_{gd} = dI_G/dV_{GD}$):

$$g_{iss} = g_{gs} + g_{gd} \tag{2-13}$$

As indicated in the curves of Fig. 2-14c and d, g_{iss} for a small-signal device is on the order of 10^{-14} A if V_{GS} is below the I_G breakpoint. In an amplifier configuration the effect of g_{gd} will be increased by the voltage gain of the amplifier. For a circuit such as that shown in Fig. 2-24, the low-frequency input conductance would be

$$g_{is} = g_{gs} + g_{gd}(1 + g_{fs}R_L) \tag{2-14}$$

(assuming that $1/g_{os} \gg R_L$).

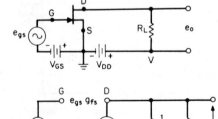

FIGURE 2-24 FET voltage amplifier and equivalent circuit (low-frequency simplified).

Even with a voltage gain of 10, g_{is} of the FET amplifier will be less than 10^{-12} mho at room temperature. Consequently, a FET has a low-frequency input resistance in excess of 10^{12} Ω, since resistance is the reciprocal of conductance.

At values of V_{DS} greater than the I_G breakpoint (see Fig. 2-14d, the dc value of I_G increases sharply. Above this corner the quantity g_{gd} becomes negative, resulting in g_{iss} going through zero and becoming negative. This occurs because I_G in this region is approximately a linear function of I_D. Since $dI_D = g_{fs}dV_{GS}$, a positive change in V_{GS} *results in an increase in I_D,* which, in turn, causes an increase in the magnitude of I_G. Since I_G is negative, dI_G is negative; therefore, g_{gd} will be negative. Fig. 2-25 is a plot of experimental data obtained from a typical JFET operating with V_{DS} greater than the I_G breakpoint. These data show that the input conductance g_{iss} is negative for this bias condition. A transition from positive g_{iss} to negative g_{iss} will occur near the knee of the I_G-versus-V_{DS} curves as was shown in Fig. 2-14d. The FET low-frequency equivalent circuit with the gate conductance components added is shown in Fig. 2-26. The g_{iss} characteristic is not normally specified in FET data sheets; however, its magnitude can be estimated from the I_G specification if provided.

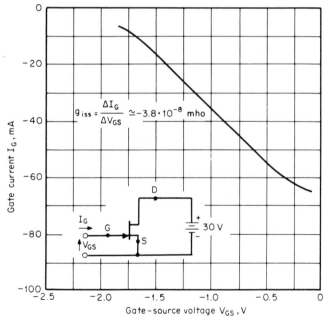

FIGURE 2-25   Gate input characteristic when V_{DS} exceeds the I_G breakpoint.

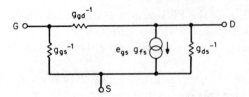

FIGURE 2-26 JFET low-frequency equivalent circuit.

ON Resistance

A depletion-mode FET, operating as a switch is considered ON when the gate-to-source voltage $V_{GS} = 0$. However, only when the drain voltage V_{DS} is well below V_p does the depletion-mode FET offer the lowest ON resistance. This is apparent from examination of Fig. 2-14a where $r_{ds(on)} = dV_{DS}/dI_D$ (slope). In the ON condition the channel conduction is a function of gate voltage and drain-source voltage. Since the slope is a function of V_{GS} we find the FET responds as a voltage-controlled resistor provided $V_{DS} \ll -V_{GS(off)}$ (Fig. 2-20). Generally when the FET is ON, V_{DS} is low; therefore, the ON resistance is specified at a low value of V_{DS}. Fig. 2-27 shows the channel conductance characteristics for n-channel and p-channel depletion-type FETs.

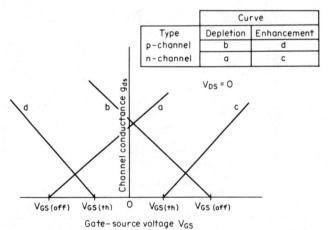

FIGURE 2-27   Channel conductance g_{ds} vs. gate-source voltage V_{GS}.

Equations for these characteristics are as follows. For n-channel depletion type,

$$r_{ds(on)} = \frac{1}{g_{fs}} = \frac{1}{K(V_{GS} - V_{GS(off)})} \qquad (2\text{-}15)$$

where $V_{GS} > V_{GS(off)}$.

For p-channel depletion type,

$$r_{ds(on)} = \frac{1}{g_{fs}} = \frac{1}{K(V_{GS(off)} - V_{GS})} \qquad (2\text{-}16)$$

where $V_{GS} < V_{GS(off)}$, and K is a constant.

Interelectrode and Parasitic Capacitances

As the operating frequency is increased, device capacitances become important parameters. For the JFET the principal capacitance is that of the gate-channel junction.

The capacitance per unit area of the p-n junction between the gate and the source and drain is a function of the junction depletion-layer thickness. The depletion layer acts as a dielectric between the p and n conducting regions. A reverse bias on the p-n junction causes the depletion layer to increase and thus the capacitance to decrease. Concepts of the junction depletion as a function of gate-to-channel voltage and of drain-to-source voltage are shown in Figs. 2-7 and 2-8. Theory indicates that the capacitance of an abrupt junction is an inverse function of the square root of the junction voltage.

$$C = \left(\frac{K}{V_{bi} + V_G} \right)^{1/2} \qquad (2\text{-}17)$$

For silicon the "built-in" space-charge voltage V_{bi} is about 0.6 V.

Abrupt junctions are not achieved in the typical FET structure. The channel may consist of an epitaxy n-type layer on a p-type substrate and with a diffused p-type top gate. The diffusion process will cause some gradation of the junctions. For a graded junction the exponent of Eq. (2-17) would be one-third instead of one-half. In a functioning circuit the analysis of gate-to-channel capacitance is complicated by the nonuniform depletion width along the channel. The gate-drain depletion width will be greater than the gate-source depletion width, thus C_{gd} usually will be lower than C_{gs}. Figure 2-28 shows the equivalent-circuit location of these

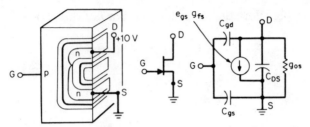

FIGURE 2-28 JFET capacitances, intermediate-frequency model.

capacitances, plus the drain-source capacitance, and gives a concept of the relative depletion-layer thickness of the drain-gate and source-gate junctions. This figure explains why the value of C_{gd} is typically less than that of C_{gs}. In some applications where V_{GD} and V_{GS} are approximately equal (such as in an analog switch or voltage-controlled resistor circuit), C_{gd} and C_{gs} may be approximately equal. The C_{ds} component is largely the device header (package) capacitance and is typically small compared to the other two components; however, at very high frequencies it must be considered.

Figure 2-29 shows the effect of gate voltage upon C_{gs} and C_{gd} for a typical JFET designed for high-frequency (450 MHz) applications. The C_{gd} curve is lower than the C_{gs} because of the 10-V V_{DS} bias (at $V_{GS} = 0$, $V_{GD} = -10$ V). These capacitances increase the input and feedback admittance and limit the wideband amplifier performance.

Because FET capacitances are sensitive to applied voltages, the circuit designer must examine the test conditions used in the device data sheet

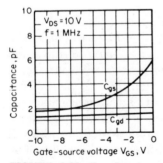

FIGURE 2-29 Junction capacitances vs. gate voltage.

specifications. Capacitances are often specified differently for different applications.

FET type	Characteristic	Max pF	Test conditions	
2N4092	C_{iss} common-source input capacitance	16	$V_{DS} = 20$, $V_{GS} = -5$	$f = 1$ MHz
2N4092	C_{rss} common-source reverse transfer capacitance	5	$V_{DS} = 0$, $V_{GS} = -20$ V	
2N4392	C_{iss}	14	$V_{DR} = 20$, $V_{GS} = 0$	$f = 1$ MHz
2N4392	C_{rss}	3.5	$V_{DS} = 0$, $V_{GS} = -7$ V	
2N4857	C_{iss}	18	$V_{DS} = 0$, $V_{GS} = -10$ V	$f = 1$ MHz
2N4857	C_{rss}	8	$V_{DS} = 0$, $V_{GS} = -10$ V	
2N5564	C_{iss}	12	$V_{DG} = 12$, $I_D = 2$ mA	$f = 1$ MHz
2N5564	C_{rss}	3	$V_{DG} = 12$, $I_D = 2$ mA	$f = 1$ MHz
2N5638	C_{iss}	10	$V_{DS} = 0$, $V_{GS} = -12$ V	$f = 1$ MHz
2N5638	C_{rss}	4		
J111	$C_{dg(off)}$	5	$V_{DS} = 0$, $V_{GS} = -10$	
	$C_{sg(off)}$	5	$V_{DS} = 0$, $V_{GS} = -10$	
	$C_{dg(on)} + C_{sg(on)}$	28	$V_{DS} = V_{GS} = 0$	

In most amplifier applications the drain-gate voltage is greater than the source-gate voltage; thus the drain-gate capacitance will be lower than the source-gate capacitance.

When using FETs as analog switches C_{ds} may play a role in establishing the OFF isolation. All the parasitic capacitances play an important role in determining the switching time of the FET. These effects are discussed more fully in Chapter 5.

Temperature effects upon gate capacitance are a function of how the device is biased. The change in depletion-layer thickness tends to give the junction capacitance a positive coefficient; however, if the bias is large (drain-gate voltage, for example), then the percentage change in depletion-layer thickness (and thus the capacitance) is small.

Switching Speed

Despite the accepted fact that FETs are voltage-controlled transistors, any change of state is accomplished by a change in charge. The FET, however, differs from the bipolar transistor in that to maintain a state does not require a continuous flow of charge. Because the FET does not require this continuous current flow we tend to forget that it is a *charge-coupled device!*

The switching speed is a direct function of charge transfer. The speed with which we are able to move the charges dictates the speed in which the FET changes state. To a lesser extent the initial turn ON and turn OFF of the JFET are a function of $V_{GS(off)}$. A low $V_{GS(off)}$ JFET will turn OFF more rapidly and ON more slowly than a JFET with a high $V_{GS(off)}$.

2-2-3 Noise Characteristics

For analysis, it is convenient to represent noise in a FET by assuming that an ideal noise-free device has two external noise sources, $\bar{e}_n$ and $\bar{i}_n$. These noise sources are chosen to have the same output as would an actual noisy FET. They both are frequency dependent within the audio-frequency noise spectrum and have the characteristics shown in Fig. 2-30. An equivalent noise circuit is shown in Fig. 2-31. The equivalent short-circuit

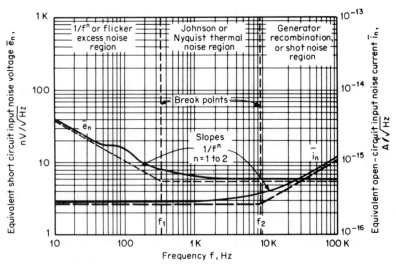

FIGURE 2-30   Characteristics of JFET noise.

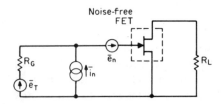

FIGURE 2-31   Equivalent JFET noise circuit.

input noise voltage, $\bar{e}^n$ (with the exception of the $1/f^n$ region), is defined as

$$\bar{e}_n = (4kTR_NB)^{1/2} \tag{2-18}$$

where $R_N \cong 0.67/g_{fs}$, the equivalent resistance for noise.

$T =$ absolute temperature in Kelvin (273 K $=$ 0°C)

$K = 1.374 \times 10^{-23}$ J/K.

$B =$ frequency range (bandwidth), Hz.

In the so-called $1/f^n$ region, $\bar{e}_n$ is expressed as

$$\bar{e}_n = \left[4KR_NB(1 + \frac{f_1}{f^n}\right]^{1/2} \tag{2-19}$$

where n varies between 1 and 2 and is device- and lot-related.

The characteristic bulge in $\bar{e}_n$ in the $1/f^n$ region has been observed to some extent in all junction FETs. This bulge is believed to be the result of generator-recombination centers. The breakpoint or corner frequency, shown as f_1 in Fig. 2-30, is lot- and device-design oriented, and varies from about 100 Hz to 1 kHz.

As indicated in Eqs. (2-18) and (2-19), $\bar{e}_n$ is inversely proportional to the square root of the transconductance of the FET ($R_N \cong 0.67/g_{fs}$). $\bar{e}_n$ can be lowered by a factor of $1/(N)^{1/2}$ if N devices with matched electrical characteristics are connected in parallel. For example, when $N = 2$ let

$$\bar{e}_{n1} = \bar{e}_{n2} \tag{2-20}$$

and let

$$g_{fs1} = g_{fs2} \tag{2-21}$$

Thus,

$$g_{fs(total)} = 2g_{fs1} \text{ or } 2g_{fs2} \tag{2-22}$$

From Eq. (2-18)

$$\bar{e}_n = \left[4KT(\frac{0.67}{2g_{fs1}})B\right]^{1/2} \tag{2-23}$$

and

$$\bar{e}_{n(total)} = \left[4KT(\frac{0.67}{2g_{fs1}})B\right]^{1/2} \tag{2-24}$$

Thus,

$$\bar{e}_{n(total)} = \frac{1}{(2)^{1/2}} \bar{e}_{n1} \tag{2-25}$$

A second way to achieve low $\bar{e}_n$ is to use a device with a large gate area. Empirically, $\bar{e}_n$ is inversely proportional to the square of the gate area, independent of g_{fs}.

The equivalent open-circuit input noise current $\bar{i}_n$, except in the shot noise region shown in Fig. 2-30, is due to thermally generated reverse current in the gate-channel junction. It is defined as

$$\bar{i}_n = (2qI_GB)^{1/2} \tag{2-26}$$

where $q = 1.602 \times 10^{-19}$ C (the magnitude of the electron charge)
   I_G = measured dc operating gate current, A
   B = bandwidth, Hz.

The expression is accurate only when the measured gate current is the result of bulk device conductance. It is possible for the measured gate current to be due to conductance stemming from contamination across the leads of the semiconductor package. At higher frequencies, as the shot-noise region shown in Fig. 2-30, $\bar{i}_n$ can be approximated as being equal to the Nyquist thermal noise current generated by a resistor:

$$\bar{i}_n = \left[\frac{4KTB}{R_p}\right]^{1/2} \tag{2-27}$$

where R_p is the real part of the gate-to-source input impedance. The breakpoint of corner frequency f_2, like f_1, in Fig. 2-30, is also lot- and device-design oriented. It usually is between 5 and 50 kHz.

Noise Figure

A noise factor F is a figure of merit of a device with respect to the resistance of a generator, shown in Fig. 2-31 as R_G. To calculate a noise factor, a source resistor R_G with a thermal noise voltage $\bar{e}_T$, is added to the circuit. A noise factor may be defined as

$$F = 1 + \frac{\text{noise power of FET referred to input}}{\text{noise power due to } R_G} \tag{2-28}$$

The thermal noise voltage across R_G is

$$\bar{e}_T = (4KTR_GB)^{1/2} \tag{2-29}$$

where $K = 1.380 \times 10^{-23}$ J/K (Boltzmann's constant)
   T = temperature, K
   B = bandwidth, Hz.

Therefore, noise power due to R_G is

$$\frac{\bar{e}_T^2}{R_G} = \frac{4KTR_GB}{R_G} = 4KTB \tag{2-30}$$

The noise power of the FET referred to the input is

$$\frac{\bar{e}_n^2}{R_G} + \bar{i}_n^2 \, R_G \qquad (2\text{-}31)$$

When expressions for the noise power of both the FET and R_G are substituted, the noise factor becomes

$$F = 1 + \frac{\bar{e}_n + \bar{i}_n^2 \, R_G^2}{4KTR_GB} \qquad (2\text{-}32)$$

A noise figure NF expressed in decibles indicates the presence of added noise power from the FET or another active device. The noise figure is always given with reference to a standard, specifically the generator resistance R_G:

$$NF = 10 \log F \qquad (2\text{-}33)$$

The noise figure of the FET is

$$NF = 10 \log \left(1 + \frac{\bar{e}_n^2 + \bar{i}_n^2 \, R_G^2}{4KTR_GB} \right) \, dB \qquad (2\text{-}34)$$

When junction FET noise is expressed in terms of the noise figure NF, an inherent disadvantage arises in that the noise figure is dependent upon the value of the generator resistance R_G. Therefore, the $\bar{e}_n$, $\bar{i}_n$ method remains as the best way to qualitatively express the noise characteristics of the FET itself.

Unlike bipolar transistors, where $\bar{e}_n$ and $\bar{i}_n$ characteristics vary directly with change in collector current I_C, similar characteristics in junction FETs will vary only slightly as drain current I_D is varied. This is true so long as the FET is biased so that the drain-gate voltage is greater than the pinchoff voltage ($V_{DG} > V_p$).

The $\bar{e}_n$ in junction FETs will be lowest when the devices are operating at $V_{GS} = 0$ ($I_D = I_{DSS}$), where transconductance g_{fs} is at its highest value. This will only be true if device dissipation is maintained very low in relation to the total dissipation capability of the FET.

The curves in Fig. 2-32 illustrate changes in $\bar{e}_n$ as the operating drain current I_D is varied. Note the lowest $\bar{e}_n$ did not occur at $V_{GS} = 0$, because of high power dissipation and a resultant rise in junction temperature at the operating point.

The optimum (lowest) $\bar{i}_n$ in depletion-mode JFETs should occur at $V_{GS} = 0$ ($I_D = I_{DSS}$). In practice, very little change will be seen in $\bar{i}_n$ when the operating point is changed, provided that the drain-gate voltage is maintained below the I_G breakpoint and power dissipation is kept at a low level. The curves in Fig. 2-33 illustrate $\bar{i}_n$ characteristics as a function of

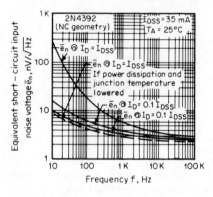

FIGURE 2-32 e_n changes vs. I_D variations.

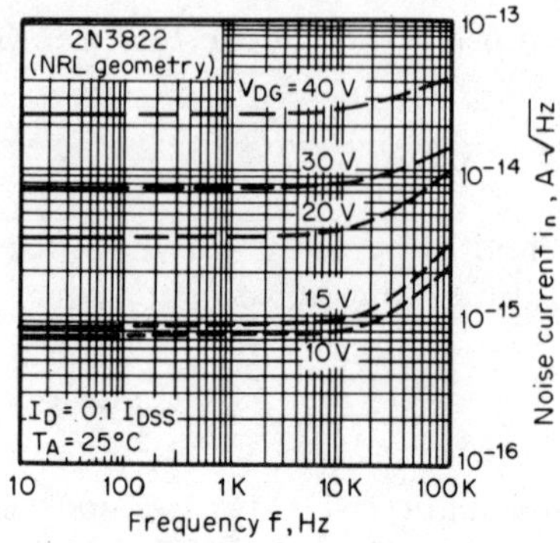

FIGURE 2-33 i_n characteristics as a function of drain-gate voltage.

drain-gate voltage at points below, on, and above the I_G breakpoint voltage.

Three equations presented earlier show that $\bar{e}_n$ and $\bar{i}_n$ are proportional to $(T)^{1/2}$. Both will be reduced if the temperature is lowered. In Eq. (2-26), $\bar{i}_n$ is proportional to $(I_G)^{1/2}$. If I_G is below its breakpoint, I_G will halve for each temperature drop of 10 to 11°C. $\bar{e}_n$ is also proportional to $(R_N)^{1/2}$, where $R_N = 0.67/g_{fs}$. Thus when g_{fs} is increased, which is typical of junction FETs operating at low temperature, $\bar{e}_n$ will also be reduced. However, g_{fs} does not continually improve as the temperature drops. Fig. 2-34 shows g_{fs} versus temperature for a silicon junction FET. Performance improves with decreasing temperature until about 100 K is reached. Below 100 K g_{fs} drops rapidly.

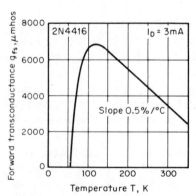

FIGURE 2-34 Effect of temperature on g_{fs}.

Test Measurements

By definition, $\bar{e}_n$ and $\bar{i}_n$ are referred to the input of the device under test. To measure $\bar{e}_n$, the test circuit shown in Fig. 2-35 is useful.

The following procedure should be used to make the $\bar{e}_n$ test:

1. Set tunable filter to required f_{low} and f_{high}. Adjust oscillator to mean center frequency

$$[f_{mean} = (f_{low} \cdot f_{high})^{1/2}]$$

2. Set V_{osc} to 100 mV with switch 1 in position 1. Compute

$$V_{in1} = 10^{-1} \times \frac{10^2}{16^6} = 10^{-5}\ \text{V} = 10\ \mu\text{V}$$

3. Measure V_{out1}. Compute overall gain as

$$A_V = \frac{V_{out1}}{V_{in1}} = \frac{V_{out1}}{10\mu\text{V}}$$

4. Set switch 1 to position 2 and measure V_{out2}. Compute V_{in2}, the equivalent short-circuit input noise voltage $\bar{e}_n$, using A_V from Step 3.

$$V_{in2} = \frac{V_{out2}}{A_V} = \bar{e}_n$$

in volts over bandwidth f_{low} to f_{high}.

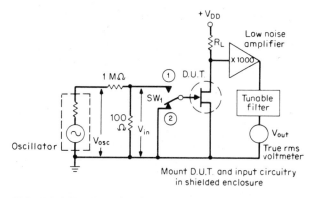

FIGURE 2-35   Test circuit to measure $\bar{e}_n$.

An alternative method of performing the above test is to use a Quan-Tech Transistor Noise Analyzer consisting of a Model 2173 Control Unit and a Model 2181 Filter. The analyzer has provision for measuring $\bar{e}_n$ and determining NF with various values of R_G in FET and bipolar devices with selectable test conditions. The measuring system has a constant gain of 10,000. The analyzer records output noise at selected frequencies between 10 and 100 Hz in the device under test, with the scale shown as the actual output divided by 10,000. This is then the output noise referred to the input. The equivalent bandwidth for testing is 1 Hz.

There are certain instances where the test circuit or the Transistor Noise Analyzer is not adequate to measure $\bar{e}_n$ at certain frequencies over certain bandwidths in the $1/f^n$ region. The rms noise over a bandwidth from f_{low} to f_{high}, where there is a $1/f^n$ characteristic over the entire range, can be computed as

$$\bar{e}_n = (e_n \text{ known})\left[f_{known} \cdot \ln\left(\frac{f_{high}}{f_{low}}\right) \right]^{1/2n} \qquad (2\text{-}35)$$

Figure 2-36 represents this equation graphically. For example, $\bar{e}_n$ known $= 70 \times 10^{-9}$ V/$\sqrt{\text{Hz}}$ at 10 Hz. How much noise is in the band from 4.5 to 5.5 Hz? The noise has a $1/f^1$ characteristic over the entire range. Thus

$$\bar{e}_n = (70 \times 10^{-9})\left[10 \cdot \ln\left(\frac{5.5}{4.5}\right) \right]^{1/2} \text{V} \qquad (2\text{-}36)$$

or

$$\bar{e}_n = 99.16 \times 10^{-9} \text{ V/}\sqrt{\text{Hz}} \ @ \ 4.975 \text{ Hz} \qquad (2\text{-}37)$$

4.975 Hz is the mean center frequency where $f_{mean} = (f_{low} \cdot f_{high})^{1/2}$.

i_n measurements are difficult to implement at best. At frequencies below f_2 in Fig. 2-40, i_n is assumed to have a constant level or "white" noise characteristic which may be correlated with gate current I_G. From Eq. (2-26) I_G is established as the measured bulk gate current. Because

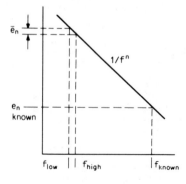

FIGURE 2-36 Computing rms noise over a bandwidth.

measured gate current I_G is the result of all conductances at the gate, the resultant gate current and the computed $\bar{i}_n$ due to bulk material can be assumed to be this value or less. The total equivalent input noise of the FET can be approximated by

$$\bar{e}_{ni}^2 = \bar{e}_T^2 + \bar{e}_n^2 + \bar{i}_n^2 \, R_G^2 \qquad\qquad 2\text{-}38$$

$\bar{e}_T$ = thermal noise of the generator resistance R_G

$\bar{e}_{ni}$ = total FET noise referred to the input

This approximation assumes that the equivalent noise voltage and the current generators vary independently. Equation (2-38) implies that i_n can be calculated if $\bar{e}_n$, $\bar{e}_T$, and total noise $\bar{e}_{ni}$ are known. The difficulty here is that in MOS or junction FETs, the R_G must be very large to detect the anticipated small value of i_n. However, when R_G is very large, $\bar{e}_T$ is much greater than $\bar{i}_n^2 \cdot R_G^2$. For example, over a 1-Hz bandwidth at 25°C, if R_G is equal to 100 MΩ, then

$$\bar{e}_T^2 = 4kTR_G = 4 \times 1.38 \times 10^{-23} \times 2.95 \times 10^2 \times 10^8$$
$$= 1.63 \times 10^{-12} \text{ V}/\sqrt{\text{Hz}}$$

Anticipated $\bar{i}_n$ is

$$\bar{i}_n \simeq 10^{-15} \text{ A}/\sqrt{\text{Hz}}$$

and

$$\bar{i}_n^2 = 10^{-30} \text{ A}/\sqrt{\text{Hz}}$$

Thus

$$\bar{i}_n^2 \cdot R_G^2 = 10^{-30} \cdot 10^{16} = 10^{-14} \text{ V}/\sqrt{\text{Hz}}$$

Therefore, $\bar{i}_n^2 \cdot R_G^2$ is much less than $\bar{e}_t^2$, which renders this method of finding i_n impractical for most common MOSFETs or junction FETs.

An improved method of measuring $\bar{i}_n^2$ is to substitute a low-loss mica capacitor for resistor R_G. The mica capacitor by definition does not have an equivalent thermal noise voltage, and thus Eq. (2-38) becomes

$$\bar{e}_{ni}^2 = \bar{e}_n^2 + \bar{i}_n^2 \cdot X_C^2 \qquad\qquad (2\text{-}39)$$

where X_C = capacitive reactance or

$$\bar{i}_n = \frac{(\bar{e}_{ni}^2 - \bar{e}_n^2)^{1/2}}{X_C} \qquad\qquad (2\text{-}40)$$

When a 10-pF mica capacitor was used in the evaluation circuit (up to a frequency of 100 Hz), a correlation of from 80 to 90 percent was obtained when compared with $\bar{i}_n$ computed from measured gate current readings.

At frequencies above 100 Hz direct computation of $\bar{i}_n$ via the capacitor method becomes unwieldy because of the rapid decrease in capacitor reactance at these frequencies.

In calculating $\bar{i}_n$ at higher frequencies, an alternative method is to measure R_p, the real part of the gate-source impedance of the FET. When R_p is measured at various frequencies, the equivalent short-circuit input noise current $\bar{i}_n$ can be computed as a function of frequency. [See Eq. (2-27).] A convenient instrument for measuring R_p is the Hewlett-Packard Type 250A RX meter or equivalent. The Type 250A RX meter can measure R_p accurately up to 200 kΩ. As is shown in Fig. 2-37, this establishes the low-frequency limit of 20 MHz for $\bar{i}_n$ computed via direct measurement of R_p for the Siliconix FET Type 2N4117A. For frequencies between 100 Hz and 20 MHz, $\bar{i}_n$ must be extrapolated, as is shown in Figs. 2-37 and 2-38. For FET types with lower R_p (such as the Siliconix 2N4393), $\bar{i}_n$ can be computed down to 2 MHz, and hence extrapolated $\bar{i}_n$ between 100 Hz and 100 kHz is more accurate.

Figure 2-39 shows representative $\bar{e}_n$, $\bar{i}_n$ curves for Siliconix JFET products. Of particular importance is the geometry which by its design governs the basic noise characteristics of product types derived from it.

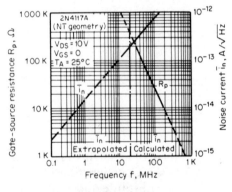

FIGURE 2-37   Low-frequency limit for calculating $\bar{i}_n$.

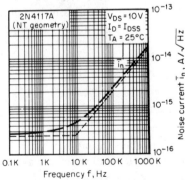

FIGURE 2-38   Extrapolated $\bar{i}_n$ vs. frequency

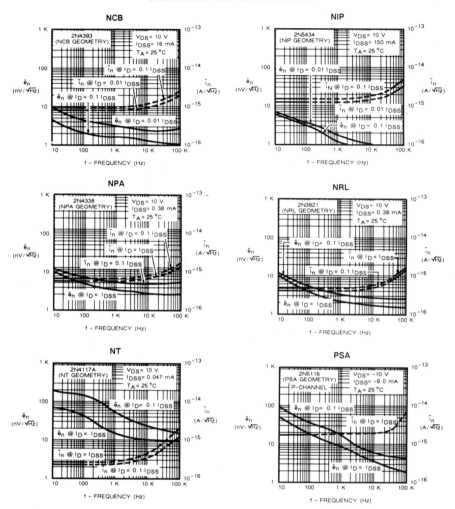

FIGURE 2-39 Noise characteristics by geometry. (Courtesy of Siliconix Incorporated.)

Conclusion

Contemporary junction FETs have noise voltages $\bar{e}_n$ equal to those found in low-noise bipolar transistors. These two device types have different operating mechanisms: the FET is voltage-actuated, while the bipolar transistor is current-actuated. Hence, FETs have an inherently lower noise current $\bar{i}_n$ and are preferred over bipolar devices in most audio-frequency applications where low-noise performance is a design requirement.

When bias points are properly selected, as described in this chapter, the excellent low-noise characteristics of high-g_{fs} junction FETs can be realized.

The curves shown in Fig. 2-39 are representatives of $\bar{e}_n$ and $\bar{i}_n$ performance of Siliconix junction FETs. Of particular importance in these curves is the process geometry by which the basic design of the FET governs the noise characteristics of product types derived from it. Readers are invited to refer to the Siliconix FET catalog for full geometry performance data and for specific part numbers stemming from the generic process geometries.

In the measurement section, it was shown that direct $\bar{e}_n$ measurements can readily be made. $\bar{i}_n$ can be guaranteed at frequencies below 100 Hz by measuring the dc operating gate current I_G. When I_G is known, $\bar{i}_n$ can be extrapolated from frequencies below 100 Hz to predict noise performance at frequencies to 100 kHz.

Another form of noise found in junction FETs is known as "popcorn" or burst noise; the term popcorn noise was originated in the hearing aid industry because of noise or level shifts which are present in input stages and which resemble the sound of corn popping. Popcorn noise is a form of random burst input noise current which remains at the same amplitude and which is confined to frequencies of 10 Hz or lower. The suitability of a FET device is dependent on the amplitude of the burst, its duration, and its repetition rate. The origins of popcorn noise are not completely identified, but it is believed to be caused by intermittent contact in aluminum-silicon interfaces and by contamination of the oxidation process.

A test circuit to measure popcorn noise in differential junction FET amplifiers is shown in Fig. 2-40. In practice, popcorn noise is evaluated on an engineering basis, not on a production-line basis. There is no apparent correlation between $1/f^n$ noise at 10 Hz and popcorn noise. However, if the amplitude of the burst is large and occurs frequently, the $1/f^n$ noise voltage ($\bar{e}_n$) is masked and difficult to evaluate at 10 Hz. The graph in Fig. 2-41 shows "moderate" burst noise observed in a group of junction FET differential amplifiers which were measured in the test circuit.

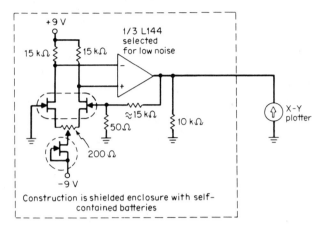

FIGURE 2-40   Test circuit to measure popcorn noise.

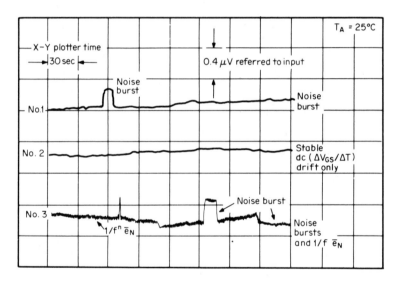

FIGURE 2-41   Popcorn noise in differential amplifiers.

2-2-4   Monolithic Duals

Differential amplifiers using FETs require close match of their electrical characteristics not only at room temperature but often over wide temperature ranges. Two-chip duals mounted in a single package cannot provide the close drift tolerances that are possible with monolithic duals. To further improve the tracking between the two FETs that comprise the monolithic dual, the metal masks of each FET often interwind as viewed in Fig. 2-42.

Monolithic duals, however, are not without problems. Chief among these are electrical interactions principally caused by the *p-n* junctions that are used to provide electrical isolation. The cross-sectional view in Fig. 2-43 identifies a potential latch-up should either gate be biased at a potential higher than the opposite source. Aside from latch-up, monolithic dual JFETs may suffer from interaction when operating at high frequencies.

Drift over temperature extremes between JFET elements of a monolithic dual can be controlled with good precision. An understanding of the effect of temperature changes upon the basic parameters of the FET

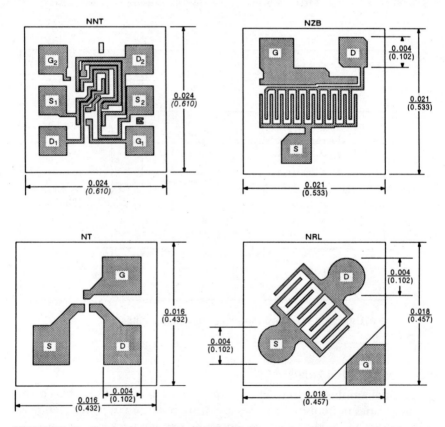

FIGURE 2-42   Comparing the monolithic dual (upper left) with the simple JFET mask, one notes the interwound elements that helps maintain uniform temperature between elements. (Courtesy of Siliconix Incorporated.)

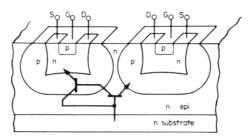

FIGURE 2-43 Cross-sectional view of junction-isolated monolithic JFET showing cause for potential SCR effect.

aids in the application of temperature-induced drift compensation. For the purpose of this discussion, temperature drift is defined as the change in V_{GS} required to maintain a constant I_D as the temperature is changed. For a typical JFET, operating with a V_{DG} greater than $-V_p$, the relationship between I_D and V_{GS} is,

$$V_{GS} = V_p \left(\frac{I_D}{I_{DSS}} \right)^{1/2} \tag{2-41}$$

V_p is a function of the zero-bias depletion-layer width at the gate-channel p-n junction. The depletion width decreases with increasing temperature, which results in an increase in channel width and therefore an increase in $-V_p$. The temperature coefficient of $-V_p$ is about 2.2 mV/°C. This widening of the channel with increasing temperature tends to increase channel conduction; however, at the same time, increasing temperature causes a *decrease* in carrier mobility and thus in channel conductivity. Carrier mobility has a negative temperature coefficient of about 0.6%/°C. The positive channel-thickness coefficient and the negative mobility coefficient have opposite effects upon the temperature characteristic of I_D and I_{DSS}. In a thick-channel device (high $V_{GS(off)}$) the mobility factor dominates and I_{DSS} has a negative temperature coefficient. In a thin-channel unit the channel-thickness effect dominates and I_{DSS} has a zero temperature coefficient. It has been shown that this particular thickness occurs in JFET devices having $V_{GS(off)}$ values of about -0.63 V. Units having *higher* $V_{GS(off)}$ values can be biased down to this critical channel dimension, so that I_D has a zero temperature drift. Mathematically the zero-drift drain current I_{DZ} is

$$I_{DZ} = I_{DSS} \left(\frac{0.63}{V_{GS(off)}} \right)^2 \tag{2-42}$$

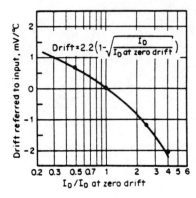

FIGURE 2-44   Drift of gate-source voltage in a single JFET drops with increasing normalized drain current. Slope of this curve is useful in determining the size of differential source resistance needed to null a given drift.

Operating at drain currents other than I_{DZ} will result in a drift in V_{GS} as given by

$$V_{GS(\text{drift})} = 2.2 \text{ mV/}°\text{C} \left[1 - \left(\frac{I_D}{I_{DZ}}\right)^{1/2} \right] \qquad (2\text{-}43)$$

A plot of this characteristic is shown in Fig. 2-44. Differentiating Eq. (2-43) with respect to I_D produces a value K for the amount of drift change per unit change of I_D:

$$\frac{dV_{GS(\text{drift})}}{dI_D} = \frac{(-1)(10^{-3})}{(I_D/I_{DZ})^{1/2}} = K \text{ V/}°\text{C} \qquad (2\text{-}44)$$

Differential JFET amplifiers and common-mode rejection is discussed in detail in Chapter 3.

2-3   THE SMALL-SIGNAL MOSFET

Unlike the JFET the MOSFET not only may exhibit depletion *and* enhancement properties but also may be constructed to provide very high mobilities, allowing maximum transconductance at low gate bias. The former MOSFET is the classic planar device whereas the latter is called DMOS (for double-diffused MOS).

2-3-1   Depletion Versus Enhancement

Common-source output and forward transconductance characteristics for two types of MOSFETs are given in Figs. 2-45 and 2-46. Figure 2-45 is for an n-channel depletion-mode device type (SD2100). The output

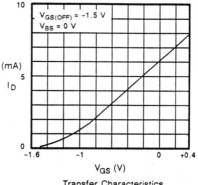

Transfer Characteristics

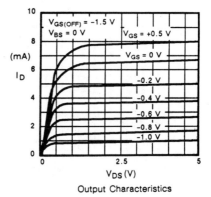

FIGURE 2-45 Output characteristics of *n*-channel depletion-mode MOSFET (SD2100) showing effects of positive gate bias.

V_{DS} (V)

Output Characteristics

and transfer characteristics indicate that the gate-source voltage may be positive or negative. Figure 2-46 shows the characteristics for an *n*-channel enhancement-mode device type (SD210DE). An important characteristic of these devices, compared with a JFET is the very low gate current. Input leakage resistance is typically greater than 10^{15} Ω.

This low input leakage characteristic makes this device useful in ultra-high input-impedance amplifiers. Caution must be exercised when handling small-signal MOSFETs as they tend to be static sensitive.

The more common planar MOSFET, built as we see in Fig. 2-47 suffers in performance as does the planar JFET simply because of the channel length. The long channel exhibits a deleterious effect upon channel conductance as shown in Fig. 2-48. A short channel FET (either junction or MOS) will tend to linearize the transfer characteristic resulting in vastly improved channel conductance by virtue of decreased degeneration. This is evident by examination and comparison of Fig. 2-49 with Fig. 2-48.

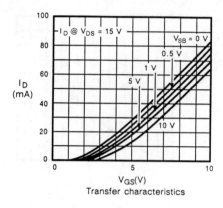

Transfer characteristics

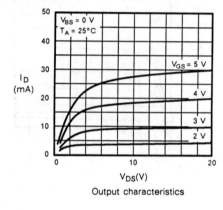

$V_{DS}(V)$

Output characteristics

FIGURE 2-46 Characteristics of enhancement-mode n-channel DMOSFET (SD210DE).

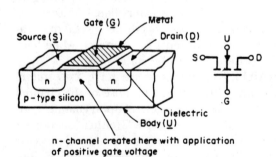

n-channel created here with application of positive gate voltage

FIGURE 2-47 n-channel MOSFET—normally OFF enhancement type.

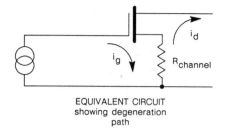

EQUIVALENT CIRCUIT
showing degeneration
path

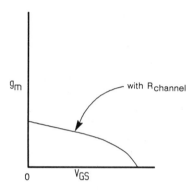

FIGURE 2-48 The degeneration
offered by a long channel exhibits
deleterious effect on transconductance.

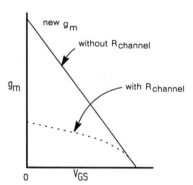

FIGURE 2-49 Increased conductance offered
by a short channel is evident in dramatic rise of
transconductance.

A dramatic improvement resulted with the introduction of the double-
diffused MOSFET (DMOS) (see Fig. 1-6) resulting in a much-improved
channel conductance. Although the DMOS offers improved frequency
response, better linearity, higher gain, and higher breakdown voltages
over the conventional planar MOSFET, we find that the general equations describing performance are equally valid for either!

2-3-2 Basic Static Parameters

Although the general equations between planar and DMOS are the same, we find slight differences between the MOSFET and the JFET. Some of these differences are influenced by substrate effects common to MOS-FETs. The JFET operates only as a depletion device whereas the MOS can be constructed so as to be either a depletion-type or an enhancement-type FET. Whereas the former performs much as does the junction FET, the latter offers a new phenomenon where drain current rises (is enhanced) with rising gate voltage.

Threshold Voltage

Threshold voltage is conveniently defined as that gate-source voltage for which a finite drain-to-source current begins to flow. Such a definition is, at best, crude, for it takes no account of channel length. A better definition, but difficult to characterize, identifies the threshold voltage as that gate-source voltage which initializes channel conduction. In MOS devices, there are what is known as surface-state traps bordering the region in and about the gate oxide. As a gate potential is gradually applied to the gate electrode the first electrons are immediately caught by these surface-state traps and, in effect, are rendered useless. Not until all the traps are filled (by our gradual increase in gate voltage) do we begin to witness conduction between drain and source. The voltage at which conduction commences is called the threshold voltage. For the n-channel enhancement-mode MOSFET this threshold voltage is positive; whereas for the p-channel enhancement-mode MOSFET, the threshold voltage is negative. Although difficult to implement in production testing, if we assume the MOSFET to have a square-law response, the threshold voltage may be extrapolated. Plotting the square-root of the drain current versus gate-source voltage at a fixed V_{DS} (well in saturation), a straight line will result that, if extrapolated to the zero-current crossing, identifies the gate threshold voltage as shown in Fig. 2-50.

For a MOSFET with a uniformly doped substrate, the threshold voltage is proportional to the square root of the applied source-to-body voltage. For DMOS where the substrate is nonuniform we see the threshold voltage behaving somewhat differently. Fig. 2-51 shows the typical threshold voltage variation as a function of the source-to-body voltage. Fig. 2-52 shows the effects of both V_{SB} and V_{GS} on $r_{DS(on)}$. It is advisable to bias the body to a voltage equal to or close to the most negative peaks of the source voltage and use a gate voltage as high as practical.

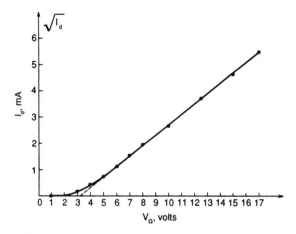

FIGURE 2-50 The true threshold voltage of a MOSFET may be determined by projecting slope of the square-root of drain current I_D to $I_D = 0$. (From Richman, *Characteristics and Operation of MOS Field-Effect Transistors*, © 1967 McGraw-Hill Book Company, New York. Used with permission.)

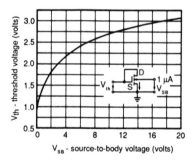

FIGURE 2-51 Threshold vs. source-to-body and gate-to-body voltage. (Courtesy of Siliconix Incorporated.)

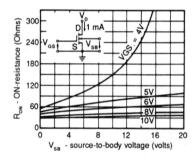

FIGURE 2-52 ON resistance vs. source-to-body and gate-to-body voltage. (Courtesy of Siliconix Incorporated.)

Gate-Source Cutoff Voltage

Planar n-channel MOSFETs fabricated on very lightly-doped p-type silicon or DMOSFETs fabricated with a n-doped channel implanted in the p-doped body region are found to operate in the depletion-mode. Like the JFET, these MOSFETs require a gate-source cutoff voltage, or a *negative* threshold voltage. The surface-state traps within the n-channel MOSFET gate oxide appear as *positive* charges. These positive charges attract equal but opposite charges in the channel. As the negative charges in the channel accumulate, channel inversion results, leading to drain-source conduction current. To halt channel conduction current the gate must be polarized sufficiently *negative* to restore the predominantly positive charge in the channel. As with the enhancement-mode MOSFET, this gate cutoff voltage is best defined at some defined drain current. For the p-channel depletion-mode MOSFET the polarities are reversed.

I_{DSS}

The depletion-mode MOSFET exhibits a zero-bias saturation drain current much the same as the JFET according to Shockley's power-law equation (Eq. 2-1).

The enhancement-mode MOSFET being normally OFF under zero-bias conditions behaves quite the opposite. At zero bias the only drain-source current flowing is the leakage current!

2-3-3   Dynamic Parameters

MOSFETs, like JFETs, find their principal use in dynamic applications where analog signals applied to the gate are amplified and extracted from the drain. Transconductance, capacitance, ON resistance, and switching characteristics are important dynamic parameters that for the MOSFET closely parallel those of the JFET. Like the JFET, Shockley's equation (Eq. 2-1) and its derivatives provides a comprehensive understanding of many of the *basic* dynamic characteristics of the MOSFET with the exception of its parasitic capacitances.

Transconductance

The transconductance and output characteristics of the MOSFET are similar to those of the JFET. The transconductance curve is shifted along the V_{GS} axis for the "normally OFF" enhancement-type device, so that a gate-source voltage of the same polarity as the drain is required to turn this unit ON. The V_{GS} at which channel conductance just starts is called the "gate-source threshold voltage" and is given by the symbol $V_{GS(th)}$. This corresponds to the $V_{GS(off)}$ of the depletion-type devices. The transconductance equations of the two types are similar, i.e.,

$$I_D = \frac{K}{2} \, (V_{GS} - V_{GS(off)})^2 \tag{2-45a}$$

where $V_{GS} > V_{GS(off)}$ (depletion type)

$$I_D = \frac{K}{2} (V_{GS} - V_{GS(th)})^2 \tag{2-45b}$$

where $V_{GS} > V_{GS(th)}$ (enhancement type). K is the device constant, which is a function of its geometry. The transconductance is

$$g_{fs} = \frac{dI_D}{dV_{GS}} = K(V_{GS} - V_{GS(off)}) \tag{2-46a}$$

where $V_{GS} > V_{GS(off)}$ (depletion type)

$$g_{fs} = \frac{dI_D}{dV_{GS}} = K \ (V_{GS} - V_{GS(th)}) \tag{2-46b}$$

where $V_{GS} > V_{GS(th)}$ (enhancement type).

Equations (2-40) and (2-41) assume that V_{DS} is greater than $(V_{GS} - V_{GS(off)})$ or $(V_{GS} - V_{GS(th)})$. As with the JFET, for the idealized device, I_D and g_{fs} saturate at a drain voltage of $(V_{GS} - V_{GS(off)})$ or $(V_{GS} - V_{GS(th)})$. Below saturation, I_D and g_{fs} are functions of both V_{GS} and V_{DS}:

$$I_D = K \left(V_p \ V_{DS} - \frac{V_{DS}^2}{2} \right) \tag{2-47}$$

where $V_p = (V_{GS} - V_{GS(th)})$ or $(V_{GS} - V_{GS(off)})$ and $V_{DS} < V_p$.

A low-frequency equivalent circuit is shown in Fig. 2-53 for the common-source configuration. The transconductance g_{fs} has been discussed above. The input conductance g_{in} is due to leakage through the thin gate oxide, device package leakage, and (if included) the input protection diode. The output conductance g_{os} has a finite value even at V_{DS} greater than "saturation" because of the decrease in effective channel length with increasing V_{DS}.

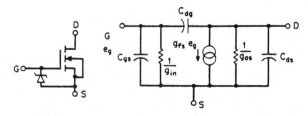

FIGURE 2-53   Low-frequency equivalent circuit for MOS-FET.

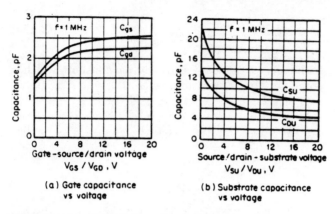

(a) Gate capacitance vs voltage

(b) Substrate capacitance vs voltage

FIGURE 2-54   MOSFET capacitance characteristics.

Capacitance

The output capacitance C_{ds} consists mostly of the drain-body n-p junction capacitance and is inversely proportional to the square-root of the drain voltage. Input capacitance C_{gs} and feedback capacitance C_{dg} are complex functions of V_{GS}, V_{DS}, substrate resistivity, threshold voltage, and overlap of the gate metal (or polysilicon) above the source and drain regions. Figure 2-54 shows short-circuit gate capacitance versus gate-source voltage for an n-channel enhancement-type MOSFET. This capacitance is related to the carrier concentration of the silicon directly beneath the gate. At a gate voltage below $V_{GS(th)}$, the "channel" region is depleted of carriers and thus C_g decreases. For V_{GS} more positive than $V_{GS(th)}$, conduction electron concentration is increased; thus, C_g increases for $V_{GS} > V_{GS(th)}$. In many devices the gate overlaps the drain region and becomes an important contributor to the feedback capacitance C_{dg}.

As with any FET, because of the nonlinear nature, these equivalent-circuit elements are functions of the operating points; thus, it is important that the bias conditions be specified when assigning values to the equivalent circuit components.

ON Resistance

The ON resistance is controlled by the electric field present across and along the channel. The oxide insulator that exists between the gate and source forms a small capacitor that accumulates charge. For an n-channel MOSFET, if the gate-to-source voltage is positive, the capacitive effect attracts electrons within the channel area immediately adjacent to the gate oxide. As V_{GS} increases the electron density in the channel will

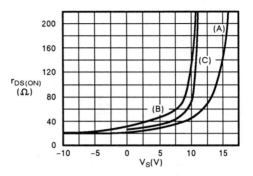

(A) $V_{BODY} = 10$ V: $V_{GATE} = 20$ V
(B) $V_{BODY} = 10$ V: $V_{GATE} = 15$ V
(C) $V_{BODY} = 0$ V: $V_{GATE} = 20$ V

FIGURE 2-55 ON **resistance of DMOS FET is a function of both the analog voltage, V_S, as well as the body and gate voltages. (Courtesy of Siliconix Incorporated.)**

exceed the hole density and inversion will result. Inversion, in this case, is when the normally p-doped channel becomes predominantly n-type.

For MOSFETs used as analog switches the ON resistance becomes a function of the amplitude and polarity of the analog voltage as well as a function of the type of MOSFET (p or n). Figure 2-55 identifies the effect of the analog voltage on ON resistance for an n-channel enhancement-mode DMOSFET.

Switching

Partially because of the low RC time constant (low $r_{DS(on)}$ and low C_{iss}) inherent in the double-diffused process, DMOSFETs are capable of extremely swift switching speeds, with turn-ON often less than a nanosecond. At these high speeds, circuit anomalies are more responsible for the observed rise and fall times than are the FETs. Using the switching test circuit shown in Fig. 2-56 the resulting switching times are offered in Table 2-1.

Although the observed turn-OFF times appear slow in comparison to the rise times, they are the result of a high load resistance R_L and the surrounding parasitic capacitances that make up both the drain node capacitances, and the circuit capacitances.

FETs, as majority-carrier transistors exhibit extremely swift turn-ON and turn-OFF switching speeds. What occurs at the drain closely follows what happens on the gate.

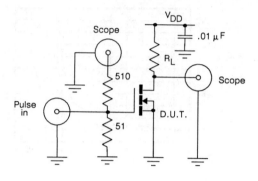

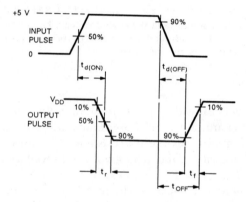

Typical Switching Waveform

FIGURE 2-56 The switching time of a DMOS is strongly influenced by circuit parameters that surround the FET. (Courtesy of Siliconix Incorporated.)

TABLE 2-1   TYPICAL SWITCHING SPEEDS

V_{DD} (V)	R_L (Ω)	$t_{d(ON)}$ (ns)	t_r (ns)	t_{OFF} (ns)
5	680	0.6	0.7	9.0
10	680	0.7	0.8	9.0
15	1K	0.9	1.0	14.0

TABLE 2-2 TYPICAL SWITCHING SPEEDS

V_{DD} (V)	R_L (Ω)	$t_{d(ON)}$ (ns)	t_r (ns)	t_{ON}* (ns)	t_{OFF} circuit (ns)
5	680	0.6	0.7	1.3	7.7
10	680	0.7	0.8	1.5	7.5
15	1K	0.9	1.0	1.9	12.1

*where $t_{ON} = t_{d(ON)} + t_r$.

We can arrive at a reasonable understanding of the apparent slow turn-OFF time in the following exercise by making one assumption: that the *DMOSFET* turns OFF as swiftly as it turns ON. Turn-ON is the sum of $t_{d(ON)}$ and t_r. If t_{OFF} for the DMOSFET equals t_{ON} for the DMOSFET, then the remaining t_{OFF} time is *circuit*-related. Table 2-1 can be revised as shown in Table 2-2.

That it *is* a circuit-related problem can be easily confirmed as follows. At $V_{DD} = 10$ V and $R_L = 680$ Ω,

Let t_{ON} for the DMOSFET (from Table 2.2) = 1.5 ns

Let t_{OFF} for the DMOSFET and circuit $= 9$ ns

The difference, the circuit, $= 7.5$ ns

Simplifying our examination, we can determine the circuit capacitance using the equation,

$$t = 2.2 \ R \ C \qquad (2\text{-}48)$$

Substituting,

$$7.5\text{E-}9 = 2.2 \times 680 \ C$$

We solve for C:

$$C = 5.01 \text{ pF}$$

In the test circuit of Fig. 2-56, we have stray and parasitic capacitances which can be itemized.

$$
\begin{aligned}
C_{\text{stray}} &= \infty \, 0.5 \text{ pF} \\
C \text{ of } R_L &= \infty \, 1.0 \text{ pF} \\
C_{\text{scope}} &= 2.0 \text{ pF} \\
\underline{C_{\text{drain}}} &= \infty \, 1.3 \text{ pF} \\
C_{\text{total}} &= 4.8 \text{ pF}
\end{aligned}
$$

These accountable capacitances closely match that calculated from Eq. (2-48). What the itemized capacitances do not show, that is included in the calculation, is the Miller capacitance.

For a drain voltage V_{DD} of $+15$ V and an R_L of 1 kΩ, we again use Eq. (2-48), this time solving for t.

$$t = 2.2 \times 1000 \times 5.01\text{E-}12 \qquad (2\text{-}49)$$

$$t = 11 \text{ ns}$$

This does not agree well with the typical t_{OFF} circuit time of Table 2.2. However, what we have not considered is the increased Miller capacitance that has slowed the fall time of the DMOSFET. We can approximate the increase in Miller capacitance if we consider the *change* in voltage gain between a circuit whose R_L was 680 Ω to what is now 1 kΩ.

$$C_{\text{Miller}} = C_{GS} + C_{GD}(A_V + 1) \qquad (2\text{-}50)$$

where

$$A_V = g_{fs} R_L \qquad (2\text{-}51)$$

From Table 2.1 the operating drain current may be determined,

$$I_D = \frac{V_{DD}}{R_L} \cong \frac{10}{680} \cong \frac{15}{1000} \cong 1.5 \text{ mA} \qquad (2\text{-}52)$$

At an $I_D = 1.5$ mA, the forward transconductance, g_{fs}, has been determined from Fig. 2-57 to be nominally 6 mmhos. Substituting into Eq. (2-51), then solving Eq. (2-50),

R_L	A_V	C_{Miller}
680	4.08	3.92 pF
1 K	6.0	4.50 pF

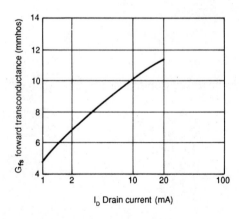

FIGURE 2-57 Effect of drain current on transconductance of a small-signal enhancement-mode DMOSFET. (Courtesy of Siliconix Incorporated.)

The difference in Miller capacitance is 0.58 pF, representing a 15% increase in C (of Eq. (2-48)), which increases t to a reasonable agreement with Table 2.2 (12.65 ns).

2-3-4   Single-Gate Versus Dual-Gate

The dual-gate MOSFET is designed to decrease the degenerative feedback effect of C_{dg}. This structure in a typical application is equivalent to a common-source amplifier followed by a common-gate amplifier. This arrangement is commonly called a "cascode" circuit. Because of the low voltage gain of the input stage, the Miller effect is greatly reduced (see Eq. (2-50)). The second gate, normally operated at or near ac (or rf) ground also provides an effective means for automatic gain control (AGC). Application of the AGC signal to gate 2 results in two advantages: 1) it provides a more remote cutoff characteristic than is achieved with a single-gate structure; and, 2) it does not affect the tuning of the input stage.

Figure 2-58a shows a section view of the dual-gate structure. This device is equivalent to two MOSFETs connected as shown in Fig. 2-58b. The common symbol is given in Fig. 2-58c. Because dual-gate MOSFETs are generally very sensitive to electrostatic damage the normal construction includes gate-protection back-to-back zener diodes, as shown in Fig. 2-58d.

The output and forward transconductance characteristics of the dual-gate MOSFET are functions of both gates. If V_{GS2} is large, the output and transfer characteristics I_D versus V_{GS1} and I_D versus V_{DS} are similar to those of a single-gate device. Under these conditions, the voltage V_{GS2} has created a low-resistance channel between drain and the internal node $S'D'$. As V_{GS2} is decreased, the transconductance characteristic curves

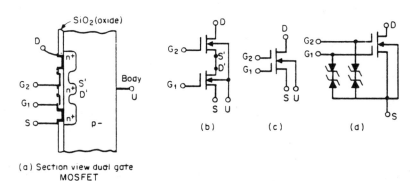

(a) Section view dual gate
     MOSFET

FIGURE 2-58   Dual-gate MOSFET—structure and symbols.

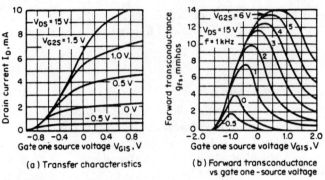

(a) Transfer characteristics

(b) Forward transconductance
vs gate one-source voltage

FIGURE 2-59   Dual-gate MOSFET characteristics.

become saturated at a drain current determined by V_{GS2} as shown in Fig. 2-59a. The resulting forward transconductance characteristics are shown in Fig. 2-59b.

2-4   MODELING

Prior to the age of the popular personal computer the closest one might come to a model was an equivalent circuit. The FET was dissected into equivalent resistors, capacitors, diodes, and generators. The Shockley model of the p-n junction was the foundation for building the FET equivalent circuit. From these equivalent circuits we were able to arrive at the many equations that are familiar to us and that are found in this chapter.

More recently SPICE modeling has become popular. However, at the time of this writing few, if any, data sheets contain sufficient information to develop a satisfactory SPICE model without consultation with the manufacturer.

2-4-1   Modeling the JFET

There are several equivalent circuits that have been proposed for the JFET. Possibly the most comprehensive and universal is that shown in Fig. 2-60. There are, of course, variations depending upon performance; the common-source, common-gate, and common-drain (source follower)—each is a slightly different variant of the model in Fig. 2-60. A high-frequency model, derived from admittance parameters, is offered in Chapter 4, and a JFET analog switch model in Chapter 5.

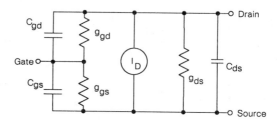

$$Y_{11} \ (b_{11}) = j \ \omega \ (C_{gs} + C_{gd})$$

$$Y_{11} \ (g_{11}) = g_{gs} + g_{gd}$$

$$Y_{22} \ (g_{22}) = g_{gd} + g_{ds}$$

$$Y_{22} \ (b_{22}) = j \ \omega \ (C_{gd} + C_{ds})$$

FIGURE 2-60   Basic JFET model.

2-4-2   SPICE Modeling

The proliferation of computer-assisted circuit design has thrust SPICE modeling into the forefront. Computer programs abound that by inputting parameter data, a reasonable model can emerge. Yet for the apparent simplicity, achieving a workable model that performs under all conditions is extremely difficult. The difficulty arises because to derive a workable model we must input the correct parameters. Furthermore, no one parameter rests without interaction with the others. So, to achieve the model we must first extract the correct parameters, which to succeed we are forced to perform successive iterations with the mandate that our results converge. Once we have successfully extracted the parameters, SPICE allows us several levels of complexity. The first level is, of course, the simple model, shown in Fig. 2-60. Each successive model required additional parameter extraction with increasing interaction.

Extraction of parameters begins with our ultimate goal of developing the *I-V* (or output characteristics) of the FET. Yet, it is not enough to arrive at a perfect fit between measured data and computational data. FETs have wide ranging specifications. Gate-source cutoff voltages, $V_{GS(off)}$, offering a 3 times spread are common. Drain saturation current, I_{DSS}, likewise may run two, perhaps three to one. Obviously the tedious manipulations that result in exact matches with measured data only holds for the FET used in the measurements. Granted the remaining FETs in that narrow distribution may also be represented by the SPICE model, but with no guarantee. Consequently, a Level 3 model, although concep-

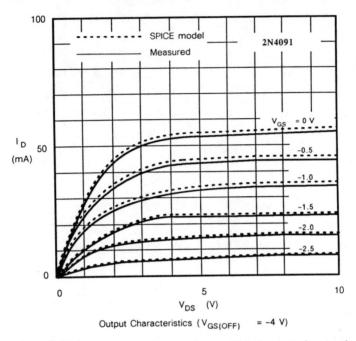

FIGURE 2-61   Comparing measured vs. modeled output characteristics of 2N4091 JFET.

tually desirable, may be quite impractical. Figure 2-61 illustrates a near-match between computational *I-V* characteristics (dashes) and measured *I-V* data (lines) for a Siliconix 2N4091 JFET. The extraction of data requires more than data sheet parameters. Sophisticated computer-programmed equipment and associated software are mandated. One popular software package for parameter extraction is called SUXES[7]

SPICE offers the user several models. The most popular being the static (or dc) model, the small-signal, dynamic model, and the noise model.

A JFET static model, derived from SPICE, is the classic FET model of Shichman and Hodge. The dc characteristics of a primitive model are determined by the SPICE parameters: VTO, BETA, LAMBDA, and IS. VTO is the gate cutoff voltage of the JFET, BETA is the gain, and IS is the gate reverse current, I_{GSS1}. BETA may be derived from the expression,

$$BETA = I_{DSS}/(VTO)**2 \qquad (2\text{-}53)$$

---

[7]K. Doganis and R. W. Dutton, "SUXES-Stanford University Extractor of Model Parameters (User's Manual)," Stanford Electronics Laboratory, Stanford University, 1982.

LAMBDA is the channel-length modulation parameter ($1/V_{DS}$) which represents that drain voltage where I_{DSS} is measured. For this simplist of models, we may extract the necessary parameters from the data sheet.

2-5 GLOSSARY OF TERMS AND ABBREVIATIONS

BV_{GSS}	Gate-source breakdown voltage
C_{bd}	Body-drain capacitance
C_{gd}	Gate-drain feedback capacitance
C_{gs}	Gate-source capacitance
C_{iss}	Common-source input capacitance
C_{oss}	Common-source output capacitance
C_{rss}	Common-source reverse transfer capacitance
C_{ds}	Drain-source capacitance
$\bar{e}_n$	Equivalent short-circuit input noise voltage
$\bar{i}_n$	Equivalent short-circuit input noise current
g_{fs}	Common-source forward transconductance
g_{fso}	Common-source forward transconductance when $V_{GS} = 0$
g_{iss}	Common-source input conductance
g_{os}	Common-source output conductance
$I_{D(off)}$	Drain cutoff current
I_D	Drain current
I_{DSS}	Saturation drain current (at $V_{GS} = 0$)
I_G	Operating gate current
I_{GSS}	Gate reverse current
$r_{ds(on)}$	Dynamic drain-source ON resistance
$r_{DS(on)}$	Static drain-source ON resistance
R_L	Drain load impedance
V_{DS}	Drain-source voltage
V_{DG}	Drain-gate voltage
V_{GS}	Gate-source voltage
V_p	Pinchoff voltage
$V_{GS(th)}$	Gate threshold voltage
$V_{GS(off)}$	Gate-source cutoff voltage

BIBLIOGRAPHY

Cobbold, Richard S. C.: *Theory and Application of Field-Effect Transistors,* John Wiley & Sons, New York, 1970.

Dacey, G. C., and **I. M. Ross:** "Unipolar Field-Effect Transistor," *Proceedings, IRE,* **41**:970–979, 1953.

Nyquist, H.: "Thermal Agitation of Electric Charge in Conductors," *Phys. Review,* **32**:110, 1928.

Radeka, V.: "Field-Effect Transistors for Charge Amplifiers, *IEEE Trans. Nucl. Sci.,* **NS-20**(1), 182–189, 1973.

Richman, Paul: *Characteristics and Operation of MOS Field-Effect Devices,* McGraw-Hill Book Co, New York, 1967.

3

LOW-FREQUENCY CIRCUITS

3-1 INTRODUCTION

FET electrical characteristics were discussed in Chapter 2, and equivalent circuit models were developed. In this chapter we will utilize these characteristics and models to develop FET amplifier configurations.

Three basic problems facing the designer of an amplifier are (1) the choice of the circuit configuration, (2) the selection of the active device type to be utilized, and (3) the setting of the bias conditions. The choices for an amplifier stage that delivers 50 W of audio power to a speaker differ from those for one which amplifies a 5-μV signal from a high-impedance source. The selection of a device and its optimum bias are interrelated; however, the approximate bias can be determined prior to the selection of a specific device type.

3-2   GROSS BOUNDARIES OF THE OPERATING REGION

For a linear amplifier there are gross boundaries on the FET output characteristics within which the operating point (bias) must be located. For reliable operation the voltage, current, and power-dissipation ratings of the FET should not be exceeded. For low distortion V_{DG} much below the I_D saturation region should be avoided. The minimum drain-voltage boundary should be such that[1]

$$V_{DG} = -V_p \tag{3-1}$$

The maximum drain voltage should be specified so that the drain-gate or drain-source breakdown voltage is not likely to be exceeded. The maximum drain current may be limited by the I_D (max) rating of the device or typically, in the case of a small-signal device, by the I_{DSS} value. I_D may be permitted to be greater than I_{DSS} if the application permits a forward bias on the gate.

The maximum power-dissipation rating results in the boundary of the maximum $V_{DS} \cdot I_D$ product. This boundary is a function of maximum operating temperature. In some cases exceeding the dissipation boundary may be permitted if the time factor is such that the rise in device temperature does not exceed the device maximum temperature rating.

The shaded area of Fig. 3-1 shows the allowed operating region. The quiescent point (zero-signal bias point) must be such that with

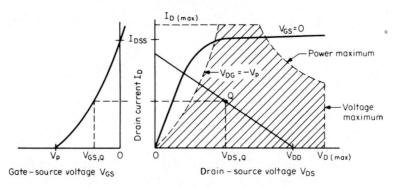

FIGURE 3-1   Gross operating boundaries for linear operation.

the maximum required output voltage and current swing, V_{DG} is not forced below $-V_p$ and I_D is not forced to zero. Further restrictions on the gross boundaries may be set by other factors such as available power-supply voltage and power consumption limits.

---

[1]We have chosen to let $V_p = V_{GS(off)}$. This differs from some writers' use of V_p to mean the value of V_{DS} at which I_D saturates. For the classical FET, with characteristics as shown in Fig. 2-14, I_D saturation occurs when $V_{DG} = -V_p$.

3-3 DESIGN EXAMPLE

We will design a FET voltage-amplifier stage and discuss some steps in choosing a suitable FET. Using this FET amplifier, we will then discuss biasing methods. For the voltage-amplifier stage shown in Fig. 3-2, make the following assumptions:

1. Voltage needed to drive the power output stage

$$e_o = 2 \text{ V rms} = 2.8 \text{ V peak}$$

2. High-frequency (-3 dB) corner $\geq$ 50 kHz

3. $R_L = 10^6$ Ω

4. $C_L = 200$ pF

5. Total available supply voltage $= 30$ V

6. $R_{gen} = 30{,}000$ Ω

7. Low-frequency corner $\leq$ 20 Hz

8. Voltage gain e_o/e_{gen} to be as high as practical

There are three amplifier configurations: the common-drain, the common-gate, and the common-source. The common-drain stage has a very high input impedance and a low output impedance; however, its voltage gain is less than 1. The common-gate stage can have a high voltage gain, but its input impedance is low. The common-source stage has both high input impedance and high voltage gain, so it is the configuration we will use for this amplifier example. We will start with the circuit of Fig. 3-2. The approximate equivalent circuits for low frequency, mid-frequency, and high frequency are shown in Fig. 3-3.

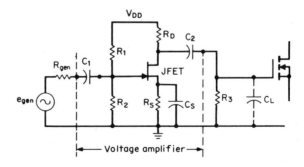

FIGURE 3-2 JFET voltage amplifier stage.

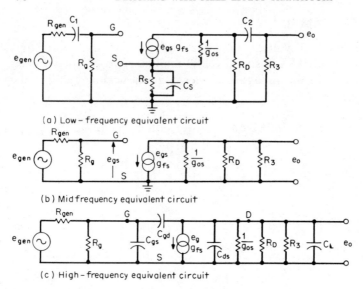

(a) Low-frequency equivalent circuit

(b) Midfrequency equivalent circuit

(c) High-frequency equivalent circuit

FIGURE 3-3 Equivalent circuit of Fig. 3-2.

3-4 ESTABLISHING THE OPERATING POINT AND BIAS METHOD

The allowed operating region has been indicated in Fig. 3-1. How do we establish the quiescent operating point and how accurately need it be maintained? Fortunately, biasing of a single-stage FET is a relatively simple matter as compared with a bipolar transistor. The input (gate) of the FET draws very little direct current, and FET dc characteristics are not as temperature-dependent as bipolar characteristics. However, parameters do vary with temperature. Also, variations will occur from device to device within the same type number. To obtain fairly uniform performance over a wide temperature range with devices of a given type, it is desirable to control I_D. When biased to a given I_D, g_{fs} varies less from unit to unit and with temperature than would be the case if it were biased with a constant V_{GS}. If the maximum possible output voltage is needed, then constant I_D biasing is the best method to use. Constant V_{GS} biasing is seldom used in low-frequency RC-coupled amplifier circuits because of the wide spread in I_{DSS} usually specified in device types.

Biasing in such a way that I_D is a controlled function of V_p may be even better than the constant-I_D method. With constant current biasing, units with a high V_p typically will have a lower g_{fs} than will units of the same device type with a low V_p. By permitting the I_D to increase for the high-V_p units, the g_{fs} can be held more nearly constant. This will result in a change in V_{DG}, which must be taken into account when considering the maximum output.

Figure 3-4 shows one method of making I_D a function of V_p. The source current is supplied through a source resistor R_S, and V_G is constant. This is sometimes called the self-bias method and is widely used for ac amplifiers.

For the forward-transfer and the output characteristic curves shown in Fig. 3-1, the source terminal was used as a reference (V_s was constant). In the circuit of Fig. 3-2, however, the gate voltage V_G is constant (assuming I_G is negligible). To study this bias situation, a different presentation of the characteristics will be used. For our design we will maintain $V_{DG} > -V_p$ so that to a first approximation I_D is affected only by V_{SG}. Figure 3-4 shows the I_D-versus-V_{SG} characteristic curves for two devices that we will consider to be our limiting units. Q_1 is the unit with maximum V_p, and Q_2 is the unit with minimum V_p. Our goal is to minimize g_{fs} variations between the limit devices. We will try to bias Q_1 so that its g_{fs} at $I_{D(1)}$ is the same as the Q_2 g_{fs} at $I_{D(2)}$. The maximum value of $I_{D(1)}$ to avoid output clipping is

$$I_{D(1)} = \frac{V_{DD} - V_G + V_P - \sqrt{2}\, e_o}{R_D} \qquad (3\text{-}2)$$

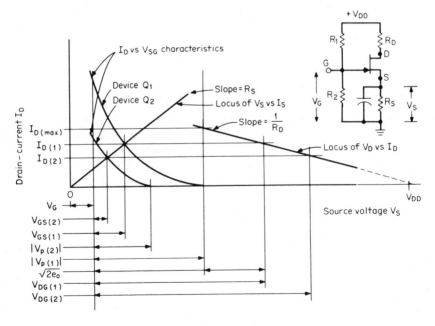

FIGURE 3-4 Bias characteristics for the amplifier circuit of Fig. 3-2.

The problem is a little more complicated now because of our desire to maintain a constant g_{fs}. It can be seen in Fig. 3-4 that for a given $I_{D(1)}$, V_G and $I_{D(2)}$ will be a function of R_S. We want $I_{D(2)}$ to be the value that results in the g_{fs} of Q_2 being the same as the g_{fs} of Q_1.

To determine $I_{D(2)}$ we make use of Eqs. (2-6) and (2-5). By letting $g_{fs(1)} = g_{fs(2)}$, we can get $I_{D(2)}$:

$$I_{D(2)} = I_{D(1)} \frac{I_{DSS(1)}}{I_{DSS(2)}} \left(\frac{V_{p(2)}}{V_{p(1)}} \right)^2 \tag{3-3}$$

where the subscripts (1) and (2) indicate devices 1 and 2.

To obtain the slope R_S, we need the values of $V_{SG(1)}$ and $V_{SG(2)}$, which may be derived from Eq. (2-1).

$$V_{GS} = V_p \left[1 - \left(\frac{I_D}{I_{DSS}} \right)^{1/2} \right] \tag{3-4}$$

Then from Fig. 3-4 we get

$$R_S = \frac{\Delta V_S}{\Delta I_S} = \frac{V_{SG(1)} - V_{SG(2)}}{I_{D(1)} - I_{D(2)}} \tag{3-5}$$

and

$$V_G = R_S I_{D(2)} - V_{SG(2)} \tag{3-6}$$

We can estimate the g_{fs} at the operating I_D with an equation derived from Eqs. (2-5) and (2-6):

$$g_{fs} = \frac{2(I_D I_{DSS})^{1/2}}{-V_p} \tag{3-7}$$

The value of R_D is needed to determine $I_{D(1)}$ and the voltage gain A_V. To maximize A_V, R_D should be high; however, an upper limit is imposed by the required frequency response. The output side of the equivalent circuit shown in Fig. 3-3 will have a high-frequency corner which will be a function of R_o. Midfrequency output voltage is

$$e_{o(MF)} = -e_g g_{fs} R'_o \tag{3-8}$$

High-frequency output is

$$e_{o(HF)} = \frac{-e_g g_{fs} R'_o}{[1 + (2\pi f C_L R'_o)^2]^{1/2}} \tag{3-9}$$

where $R'_o = \dfrac{1}{1/R_D + 1/R_G + g_{os}}$ (3-10)

The fall-off in output at high frequency resulting from C_L is obtained by the ratio of Eq. (3-9) to Eq. (3-8):

$$\frac{e_{o(\text{HF})}}{e_{o(\text{MF})}} = \frac{1}{[1 + (2\pi f C_L R_o')^2]^{1/2}} \qquad (3\text{-}11)$$

(assuming e_g is constant).

3-5 INPUT CAPACITANCE

On the input side of the high-frequency equivalent circuit, the capacitance C_{in} will cause e_g to decrease as frequency is increased. The equation to determine this loss is similiar to Eq. (3-11):

$$\frac{e_{g(\text{HF})}}{e_{g(\text{MF})}} = \frac{1}{[1 + (2\pi f C_{\text{in}} R_G')^2]^{1/2}} \qquad (3\text{-}12)$$

where R_G' is the parallel combination of R_{gen}, R_1, and R_2, and

$$C_{\text{in}} = C_{gs} + C_{gd}(1 - A_V) \qquad (3\text{-}13)$$

C_{gs} and C_{gd} are functions of V_{GS} and V_{GD}. Therefore we must determine bias conditions as well as A_V to determine C_{in}.

Junction capacitance C_j as a function of voltage has the form

$$C_j = C_o\left[1 + \left(\frac{V_j}{V_{bi}}\right)^k\right]^{-1} \qquad (3\text{-}14)$$

If C_j is known at one bias voltage $V_{j(1)}$, it can be determined at another bias voltage $V_{j(2)}$ by

$$C_{j(2)} = C_{j(1)}\left(\frac{V_{bi} - V_{j(1)}}{V_{bi} - V_{j(2)}}\right)^k \qquad (3\text{-}15)$$

For an abrupt junction, $k = 0.5$; for a linearly graded junction, $k = 0.333$. Most JFET gate junctions fall between these two conditions, so we will use $k = 0.4$. For silicon the "built-in" space-charge voltage V_{bi} is about 0.6 V. The FET specification sheet will normally give the bias condition for the C_{gs} and C_{gd} specs. For the amplifier, V_{GS} is given by Eq. (3-4). By observation from Fig. 3-4 we can get

$$-V_{GD} = V_{DD} - V_G - I_D R_D \qquad (3\text{-}16)$$

3-6 CHOOSING THE FET—A DESIGN EXAMPLE

We will use these equations in the design of the amplifier of Fig. 3-2. A point that should be emphasized is that not all JFET device types conform to the equations presented in Chapter 2 and in this chapter.

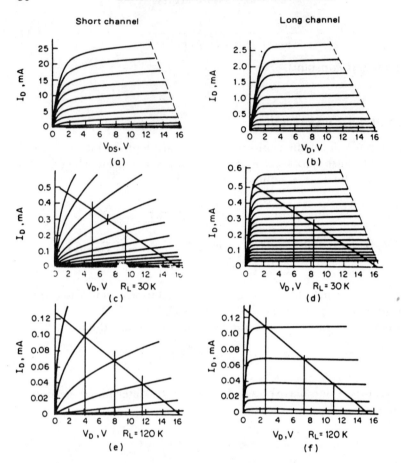

FIGURE 3-5   Comparison of short-channel JFET with long-channel JFET.

After a part number has been selected for the amplifier, its characteristics should be checked with the design operating conditions, for conformance with the design goals. For example, the output characteristics of two different FET geometries are given in Fig. 3-5. The active chip areas of these two geometries are approximately equal, and their $V_{GS(off)}$ values are about the same. It can be seen, however, that the short-channel geometry has an I_{DSS} about 10 times that of the long-channel geometry. This is due to the difference in the channel length and width of the two types, the shorter length of the short channel resulting in a higher I_{DSS} and g_{fs} per unit of channel width. It also results in a higher g_{fso}/C_{gd} ratio, which is an important factor for high-frequency amplifiers and for low-ON-resistance analog switches. The long-channel geometry, on the other hand, is good for low-noise performance at low frequencies where its relatively low g_{fs}/C_{gd} ratio is not so important as it is for high-frequency devices. Figure 3-5 compares the output characteristics of these two devices at various I_D levels.

The short-channel effect is very noticeable at low-current-bias conditions. The high g_{os} is apparent. It can also be seen, however, that when biased for the same I_D condition, the g_{fs} is much higher for the short channel than for the long channel. Even with the higher g_{os}, the A_V [Eq. (2-11)] for the short-channel unit can be greater than for the long-channel unit. These characteristics should point out the importance of not taking for granted all equation simplifications. Neglecting g_{os} to arrive at a simplified equation for A_V [Eq. (2-12)] would cause little error for the long-channel geometry with the load lines shown in Fig. 3-5; however, it would result in a large error with the short-channel geometry.

Deciding what FET type to use for a given application can be a problem. There are several hundred part numbers listed in a typical FET catalog. Fortunately almost all these numbered types are made from just a few basic production designs. These designs are referred to as "geometries." Within a given geometry the I_{DSS} and $V_{GS(off)}$ are related approximately as indicated by Eqs. (2-1) through (2-4). The major difference between devices within a given geometry is the channel thickness, usually controlled by an epitaxial-layer thickness, ion implant depth, diffusion time, or a combination of these. Figure 3-6 shows the I_{DSS}

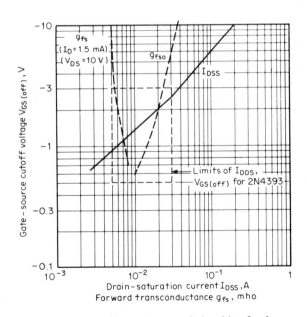

FIGURE 3-6 I_{DSS} and g_{fs} vs. $V_{GS(off)}$ relationship of a short-channel geometry (2N4393).

versus $V_{GS(off)}$ relationship of devices made from Siliconix Inc.'s "short-channel" geometry. About 40 "standard" part numbers are characterized from this one geometry.

For our amplifier example we will utilize a "JFET geometry selector guide," reproduced in Fig. 3-7 for convenience. The curves indicate a V_p range from about 0.1 to 10 V. With our knowledge that higher A_V can be achieved with lower V_o units (provided I_D is adequate), we tend to favor these units for our design. Since 3 V is about the middle of the V_p range indicated, we will use this as a tentative value.

Figure 3-4 shows the bias characteristics for the amplifier. The bias current $I_{D(1)}$ can be calculated with Eq. (3-2); however, we must first set values for R_D and V_G. Our amplifier design called for e_o to be 2 V rms, and we have assumed $V_p = -3$ V. The upper limit for R_D is determined by the maximum permissible fall-off at high frequency as indicated by Eq. (3-11). The design goal called for less than 3 dB loss at 50

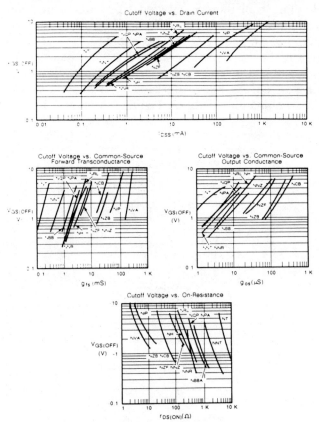

FIGURE 3-7   JFET geometry selection guide. (Courtesy of Siliconix Incorporated.)

kHz. There will be some loss at the input due to C_{in}; however, we can't determine C_{in} until we know the device we are going to use and the value of R_D. We will assume for the moment a 1.5-dB loss at the input and a 1.5-dB loss at the output.

A loss of 1.5 dB is a voltage ratio of 0.841. Using this value we use Eq. (3-11) to obtain a value for R_o:

$$R_o = \frac{[(e_{o(MF)}/e_{o(FH)})^2 - 1]^{1/2}}{2\pi f C_L} \tag{3-17}$$

With $f = 50$ kHz, $C_L = 200$ pF, and the 1.5-dB loss indicated above, we get with Eq. (3-17) a value of $R_o = 10{,}239$ Ω. R_o is the parallel combination of R_D, R_L, and $1/g_{os}$. In our example $R_L = 1$ MΩ, so it can be neglected. Figure 3-5 shows that g_{os} is a function of the FET geometry, so we cannot get its value until we select the device type. For the moment we will assume that $1/g_{os} \gg R_L$ and let $R_L = R_o$. We will modify this after we determine the approximate g_{os} for the selected device type.

The best value for V_G will be a function of the geometry; for now we will assume a value of 3 V so that we can proceed with estimating an $I_{D(1)}$ value. Using Eq. (3-2) we get

$$I_{D(1)} = \frac{30 - 3 - 3 - 2.8}{10{,}239} = 2 \text{ mA}$$

Minimum I_{DSS} of the selected FET should allow for a peak signal current above the $I_{D(1)}$ value.

$$I_{D(max)} = I_{D(1)} + \frac{\sqrt{2}\, e_o}{R_D} \tag{3-18}$$

If $I_{DSS(min)}$ is lower than $I_{D(max)}$, then the gate will become forward-biased on positive input peaks. Typically I_{DSS} will decrease as temperature increases, so this must be taken into account. The maximum temperature coefficient for I_{DSS} is approximately $e^{-(0.006)\Delta T}$. If we assume a maximum operating temperature of 75°C, then the 25°C value of I_{DSS} should be

$$I_{DSS(25°)} = I_{DSS(75°)}e^{-(0.006)(25-75)} = 1.35 I_{DSS(75°)}$$

With $I_{D(1)} = 2$ mA, $I_{D(max)}$ per Eq. (3-18) will be

$$I_{D(max)} = 2 + \frac{(1.414)(2)}{10.239} = 2.28 \text{ mA}$$

and $I_{DSS(25°)} = (1.351)(2.28) = 3.08$ mA (min).

Looking at the selector guide in Fig. 3-7 we tabulate in Table 3-1 FET geometries having $I_{DSS} > 3$ mA. We have added to this table the approximate V_p, BV_{GSS}, and capacitance values.

TABLE 3-1   SELECTED FET GEOMETRIES HAVING $I_{DSS} > 3$ mA

Siliconix geometry	Approximate $-V_p$ for $I_{DDS} = 3$ mA	BV_{GSS} range	At $V_{DS} = 10$ V and $V_{GS} = 0$	
			$C_{iss} - C_{rss}$	C_{rss}
NPA	2.3	>40	3.5	1.6
NRL	1.5	>40	4	1
NH	1.3	30–40	2	0.6
NZF	0.9	30–40	5	1
NCB	<0.5	>40	10	3
NZB	<<0.5	25–35	8	2
NVA	<0.5	25–35	160	—

High gain at midfrequency could be achieved with the NVA geometry; however, because of its large capacitance values, the high gain would probably have to be sacrificed to get the desired high-frequency corner. Also there are no standard devices from NVA geometry having a V_{DG} rating of 30 V or greater, so we will not evaluate this geometry further.

The NCB geometry is next in line with respect to I_{DSS}/V_p ratio. It has a high BV_{GSS} and reasonably low C_{iss} and C_{rss}. There are several devices using this geometry that have low V_p specs of -0.5 to -3 V. We will select the 2N4393 for further evaluation.

The NZF geometry has lower C_{iss} and C_{rss} than the NCB; however, there are no standard device types with BV_{GSS} of 30 V or greater.

The low-capacity NH geometry should be further evaluated and compared with the device selected from the NCB geometry. We find that the FET type 2N4416 has a 5-mA I_{DSS} minimum, and also has a 30-V BV_{GSS} minimum, so we select it for evaluation. The other geometries in Table 3-1 have higher V_p and higher C_{iss} than the NH, so they will not perform as well.

Table 3-2 is used to list the characteristics of the 2N4416 and the calculated amplifier-component values. In the "Note" column we indicate whether the listed characteristic is a data-sheet limit, a value estimated from the data-book curves, or a calculated value. We have a column for the high-V_p unit (Q_1 in Fig. 3-4) and a column for the low-V_p unit (Q_2 in Fig. 3-4). The performance curves will aid us in estimating the interrelation of the various characteristics. For example, we note that the typical $I_{DSS(max)}$ unit of the 2N4416 (NH geometry) has a V_p of about -4.8 V instead of the -6-V limit permitted by the specification sheet.

Also the typical $V_{p(min)}$ unit has an I_{DSS} of about 6 mA instead of the data-sheet limit of 5 mA.

TABLE 3-2 DEVICE TYPE 2N4416 (SILICONIX NH GEOMETRY)

Characteristic	Value for Q_1 (high V_p)	Value for Q_2 (low V_p)	Unit	Note
I_{DSS}	15	5	mA	1
V_p	6	2.5	V	1
C_{gs}	3.2	3.2	pF	$V_{os} = 0$
C_{gd}	0.8	0.8	pF	$V_{gd} = -15$ V
I_{DSS}	15		mA	1
I_{DSS}		6	mA	2
V_p	4.8		V	2
V_p		2.5	V	1
g_{os}	0.013	0.007	mmho	2
R_D	12	12	kΩ	Eq. (3-17)
R_o	10.4	11	kΩ	Eq. (3-10)
I_D	1.5	1.02	mA	Eqs. (3-2), (3-3)
V_{SG}	3.28	1.47	V	Eq. (3-4)
R_S	3.77	3.77	kΩ	Eq. (3-5)
V_G	2.38	2.38	V	Eq. (3-6)
g_{fs}	3	3	mmho	2
e_g/e_g	31.2	33		$g_{fs}R_o$
V_{DG}	9.62	15.38	V	Eq. (3-16)
C_{gs}	1.55	1.95	pF	Eq. (3-15)
C_{gd}	0.95	0.8	pF	Eq. (3-15)
$e_{o(HF)}/e_{o(MF)}$	0.837	0.823	—	Eq. (3-11)
$e_{g(HF)}/e_{g(MF)}$	0.957	0.965	—	Eq. (3-11)
$e_{o(HF)}/e_{g(MF)}$	0.801	0.794	—	
$20 \log [e_{o(HF)}/e_{g(MF)}]$	−1.93	−2.0	dB	
C_1	0.03		μF	Eq. (3-20)
C_2	0.03		μF	Eq. (3-20)
C_3	33		μF	Eq. (3-20)
R_1	12		mΩ	Eq. (3-21)
R_2	1		mΩ	Eq. (3-21)

NOTES:
1. Data-sheet limit.
2. Estimated from curves.
3. Calculated with equation.

In calculating $I_{D(1)}$ we will use -6 V for V_p because this will ensure that a data-sheet-limit unit will not result in output clipping at maximum e_o. For calculating $I_{D(2)}$, however, we will use the relationships indicated by the performance curve. This should result in better selection of the proper R_S for g_{fs} control.

We proceed with the calculations and fill in Table 3-2. In arriving at the first approximation for $I_{D(1)}$, g_{os} was neglected. Now that we have I_D, we can estimate from the performance curves g_{os} for the two devices. For the NH with high V_p we get $g_{os} = 0.013$ mmho.

The calculated value of g_{fs} is 1.98 mmhos. The g_{fs} versus I_D performance curve in the data book indicates that g_{fs} is closer to 3 mmhos for this device. The error in the equation is probably due to the short-channel effect in the NH geometry. We used the higher of the two values in Table 3-2.

The midfrequency gain (e_o/e_g) is slightly higher for the low-V_p unit because the lower I_D gives a lower g_{os}. This could be compensated by lowering $I_{D(2)}$ to reduce $g_{fs(2)}$.

The high-frequency loss is only 2 dB because the C_{in} loss is less than we initially allowed for. Figure 3-3a shows the low-frequency equivalent circuit. Response will be determined by the input and output coupling capacitors, and by the source-bypass capacitor. The low-frequency transfer function of a simple RC high-pass filter is

$$K_{LF} = \frac{1}{\{1 + [1/(2\pi fCR)]^2\}^{1/2}} \tag{3-19}$$

We have three RC networks to consider: (1) the input, consisting of C_1, R_{gen}, and R_g'; (2) the output, consisting of C_2, R_D, g_{os}, and R_3; and (3) the source circuit, consisting of C_S, R_S, and g_{fs}.

We let the attenuation resulting from these three RC networks be -3 dB, i.e., the product of the three K_{LF} factors be 0.708. Since the input and output R_S are much larger than the equivalent source resistance, C_1 and C_2 can be much smaller than C_S. If we let K_{LF} associated with C_1 and C_2 each $= 0.95$, then K_{LF} associated with C_S will be

$$K_S = \frac{0.708}{(0.95)(0.95)} = 0.784$$

where K_S is low-frequency attenuation resulting from imperfect source bypassing. For both the input and the output, R is approximately 10^6 Ω. Assuming $K = 0.95$ at $f = 20$ Hz,

$$C = [2\pi fR(K_{LF}^{-2} - 1)^{1/2}]^{-1} \tag{3-20}$$

Using the above values we get C_1 and $C_2 = 0.0242$ μF. We will use 0.03 μF.

For the source circuit, the equivalent R is R_S in parallel with g_{fs}^{-1}. Using Eq. (3-20) and letting $K = 0.784$, we get $C_S = 33$ μF. The voltage across C_S will not exceed $R_S/(R_D + R_S)$ (V_{DD}). A 10-V rating for C_S is used.

We have assumed the parallel combination of R_{G1} and $R_{G2} = 10^6$ Ω. The ratio is set by V_{DD} and V_G.

$$\frac{R_{G2}}{R_{G1} + R_{G2}} = \frac{V_G}{V_{DD}} \tag{3-21}$$

Since we want

$$\frac{R_{G2}}{R_{G1} + R_{G2}} = 10^6 \ \Omega$$

and

$$\frac{R_{G2}}{R_{G1} + R_{G2}} = \frac{-V_G}{V_{DD}} = \frac{2.38}{30}$$

$$R_{G(1)} = (10^6)\left(\frac{30}{2.38}\right) = 12.6 \ \text{M}\Omega$$

Use 12 MΩ. Then

$$R_{G2} = \frac{2.38}{30} \, (R_{G1} + R_{G2})$$

$$R_{G2} = \frac{2.38}{30} R_{G1}\left(1 - \frac{2.38}{30}\right)^{-1} = 1.03 \ \text{M}\Omega$$

Use 1 MΩ.

The completed amplifier design is shown in Fig. 3-8.

3-7 CONSTANT CURRENT-SOURCE BIAS

Replacing R_S in Fig. 3-8 with a current source results in the circuit and bias conditions shown in Fig. 3-9. The value of the current source I_S determines the value of I_D. If I_S is constant, then $I_{D(1)} = I_{D(2)}$. I_D does not change from device to device. From a practical standpoint a FET current-limited diode, such as the Siliconix type CR100, can be used. By setting V_G to $+ 2$ V, the CR100 operates well into its saturated region, so that its effective output resistance is greater than 180 kΩ. Hence, a V_{SG} change of 2 V results in a less than 12-μA change in I_D. However, we must take into account the current-source tolerance. If we assume

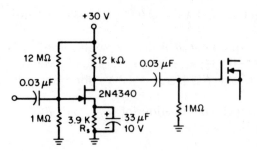

FIGURE 3-8 Completed amplifier schematic.

a ±10 percent tolerance, for example, then the I_{DQ} range would be 0.9 to 1.1 mA. As with the source resistance example, the current source must be bypassed with C_S of essentially the same value.

An advantage of the constant current source over the large source resistor is the improvement in current stability. The drain current is essentially independent of the I_{DSS} and V_p of the FET (within limits). As shown in Fig. 3.9, $I_{D(1)}$ and $I_{D(2)}$ are equal to the current of the regulator CR100. With CR100 limits of 0.9 and 1.1 mA, V_D would be

$$V_{D(high)} = 30 \text{ V} - (0.9 \text{ mA})(12 \text{ k}\Omega) = 19.2 \text{ V}$$
$$V_{D(low)} = 30 \text{ V} - (1.1 \text{ mA})(12 \text{ k}\Omega) = 16.8 \text{ V}$$

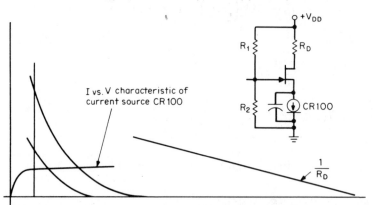

FIGURE 3-9 Characteristics with constant current bias.

Maximum output voltage without clipping would be

$$\ell_{o(high) \ max} = 30 - 19.2 = 10.8 \text{ V}$$
$$19.2 - 6 = 13.6 \text{ V}$$
$$\ell_{o(low) \ max} = 30 - 16.8 = 13.2 \text{ V}$$
$$16.8 - 6 = 10.2 \text{ V}$$

For the worst case,

$$\ell_{o(peak)} = 10.2 \text{ V}$$

An important application of the current regulator (or "current source") is in differential amplifiers. The high output resistance of the current source enhances the common-mode rejection.

3-8 CASCODE CIRCUITS

The term "Cascode" describes an amplifier stage consisting of a common-source (or emitter or cathode) stage, followed by a common-gate (or base or grid) stage. This circuit configuration was originally developed to achieve improved high-frequency performance in vacuum tube amplifiers. The Cascode configuration not only improved the output-to-input isolation; it also greatly improved the high-frequency amplifier stability by reducing the feedback capacitance within the amplifier stage. In this section Cascode circuit advantages in dc and low-frequency amplifier circuits are covered. High-frequency circuits are discussed in Chapter 4.

3-8-1 Increasing Junction-FET-Amplifier Input Resistance by Using Cascode Circuits

The gate-leakage current I_G as a function of drain-to-gate voltage V_{DG} discussed in Chapter 2 increases rapidly once the I_G breakpoint voltage is exceeded (Fig. 2-14d). Cascode circuits are useful in amplifier applications requiring high input resistance over a wide input-voltage range; for example, source followers and differential amplifiers.

3-8-2 Practical FET Cascode Circuits

Operation at V_{DG} values below the I_G breakpoint is desirable for amplifier applications requiring very high input resistance. If a large input-voltage range is required, the circuits shown in Fig. 3-10b and d have been found to be useful. The circuits maintain a low V_{DG} on the *input* FET Q_1; thus, the breakpoint of Q_1 is not encountered, and the I_3 component is undetectable. Gate current of Q_2 increases as *its* I_G breakpoint is exceeded; however, it does not contribute to input current and remains well below 0.1 percent of I_D until avalanche breakdown is approached. The curves of Fig. 3-10 compare the I_G versus V_{DG} characteristics of the circuits shown.

Another advantage of the Cascode circuit is a reduction in input capacitance due to a decrease in the Miller effect. Figure 3-10 shows C_{in} versus V_{DG} for the amplifier circuits. The great reduction of C_{in} of the source follower of Fig. 3-10d is caused by the "bootstrapping" of the Q_1 drain to its source. Thus, both the drain and the source of Q_1 are forced to "follow" the gate, greatly reducing the effect of both C_{gd} and C_{gs} upon C_{in}.

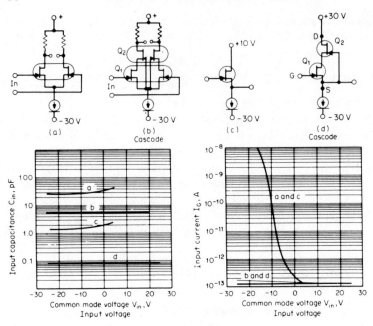

FIGURE 3-10   Cascode connections to reduce I_G and C_{in}.

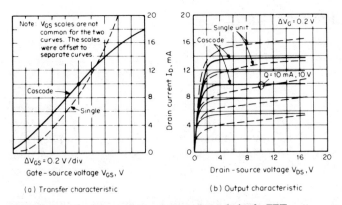

FIGURE 3-11   Comparison of Cascode and single FET.

Figure 3-11 shows the transconductance and static output characteristics of the Cascode circuit for comparison; the broken-line curves indicate the characteristics of the single device Q_1. At the selected operating point ($I_D = 10$ mA) the transconductance is approximately the same

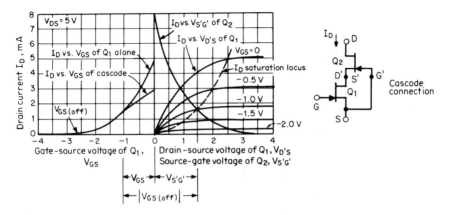

FIGURE 3-12 Characteristics of Cascode circuits.

as for the single device. The output conductance, however, is much lower for the Cascode connection.

Figure 3-12 will aid in understanding the Cascode characteristics. Note that V_{GS} of $Q_2 = V_{DS}$ of Q_1. The operating point for Q_1 will be at the intersection of the I_D-versus-V_{SG} curve of Q_2. The broken-line curve shows the approximate V_{DS} at which I_D saturation occurs for Q_1. The optimum operating point for Q_1 is near its I_D saturation. At lower values, transconductance begins to drop due to an increase in g_{ds}. Higher values will result in some increase in I_G.

The voltage at which drain current saturates is

$$V_{DS} = V_{GS} - V_{GS(off)} \tag{3-22}$$

$$V_{DG} = -V_{GS(off)} \tag{3-23}$$

To keep both Q_1 and Q_2 operating in drain-current saturation, the minimum drain-gate voltage of the Cascode connection should be

$$V_{DG(min)} = -V_{GS(off)(1)} - V_{GS(off)(2)} \tag{3-24}$$

If Q_1 and Q_2 are matched units, then

$$V_{DG(min)} = -2V_{GS(off)} \tag{3-25}$$

The Cascode FET pair, as shown in Fig. 3-13a, may be regarded as an active two-port. The midfrequency equivalent circuit shown in Fig. 3-13b may be reduced to the simplified circuit shown in Fig. 3-13c. The feedback capacitance C_{gd} and the output conductance g_{os} are reduced by the ratio g_{ds}/g_{fs}.

The voltage gain (A_V) of the *input* stage of the Cascode is approxi-

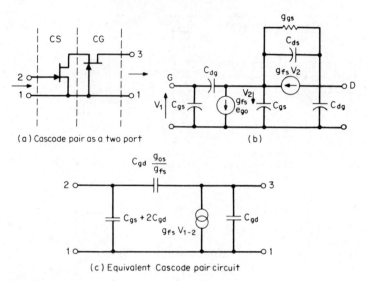

(a) Cascode pair as a two port

(b)

(c) Equivalent Cascode pair circuit

FIGURE 3-13   Cascode equivalent circuit.

mately -1; therefore, the input capacitance (including the Miller effect) is

$$C_{in} = C_{gs} + 2C_{gd} \qquad (3\text{-}26)$$

where C_{gd} and C_{gs} are values for the *input* FET.

Output capacitance is approximately C_{gd} of the common-gate output stage. Forward transconductance g_{fs} will be equal to the input-stage transconductance.

At a given drain current level the effective "bottoming" voltage (minimum V_{DG}) for the Cascode is about twice that for a single device; thus the power efficiency is lower. For small-signal amplifiers, which are the major application of the Cascode, this is not important. A commonly used Cascode configuration for FM and TV front ends is the dual-gate MOSFET, which in effect is an integrated Cascode pair (shown in Fig. 3-14). The 3N201 is an example of such a device. Its characteristics are discussed in Chapter 4.

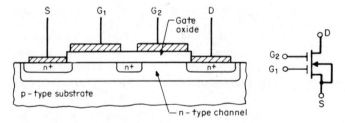

FIGURE 3-14   Dual-gate MOSFET.

The reduction of I_G and C_{in} by the use of the Cascode connection can be used to advantage in differential amplifiers. The common-mode voltage range can be greatly increased and the input bias current reduced. The Cascode differential amplifier is further discussed in Sec. 3-10.

3-9 SOURCE FOLLOWER (COMMON-DRAIN AMPLIFIER)

Too little knowledge of biasing methods for FET amplifiers sometimes keeps engineers from making maximum use of FETs in circuit designs. The common-drain amplifier, or source follower, is a particularly valuable configuration. Its high input impedance and low output impedance make it very useful for impedance transformations between FETs and bipolar transistors.

Figure 3-15 shows 10 circuits, which represent virtually every source-follower configuration. Understanding these circuits will help the designer obtain consistent circuit performance despite wide device variations.

There are two basic connections for source followers: with and without gate feedback. Each connection comes in several variations. In Fig. 3-15, circuits a through e have no gate feedback; their input resistances are approximately equal to R_G. Circuits f through k employ feedback to their gates to increase the input resistance above R_G.

Before getting into the details of bias-circuit design, note several general observations that can be made.

○ Circuits a, d, and f can accept only positive and small negative signals because these circuits have their source resistors connected to ground. A signal equal to $V_{GS(off)}$ will cut off I_D. The other circuits can handle large positive and negative signals limited only by the available supply voltages and device breakdown voltage.

○ Circuits c, d, e, h, j, and k employ current sources to improve drain-current (I_D) stability and increase gain.

○ Circuits d, e, and k employ FETs as current sources. In circuit d, Q_2 must have a lower cutoff voltage $V_{GS(off)}$ and a lower zero gate-voltage drain current I_{DSS} than Q_1.

○ Circuits e, g, h, and k employ a source resistor R_S which may be selected to set the quiescent output voltage equal to zero.

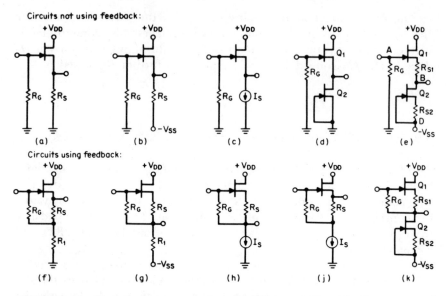

FIGURE 3-15 Virtually every practical source-follower configuration is represented in this collection of 10 circuits. Configurations (a) through (e) do not employ gate feedback; the corresponding ones in the bottom row do. (Reprinted with permission from J. S. Sherwin, "Build Better Source Followers 10 Ways," *Electronic Design*, Vol. 18, No. 12, copyright Hayden Publishing Co., Inc., 1970.)

○ Circuits e and k use matched FETs. R_S is selected to set I_D near the specified low-drift operating current. The input-output offset is zero if $R_{S1} = R_{S2}$ and the FETs are matched.

3-9-1 Biasing without Feedback Is Simple

In Fig. 3-15, the no-feedback circuits a through e use simple biasing techniques. Circuit a is a self-bias configuration; the voltage drop across R_S biases the gate (which draws essentially zero current) through resistor R_G. Since no gate-to-source voltage V_{GS} can be developed when $I_D = 0$, the self-bias load line passes through the origin as shown in Fig. 3-16. For the 2N4339 FET, whose limiting transfer characteristics are used as an example, the quiescent drain current is seen to lie between about 0.25 and 0.55 mA when a 1-kΩ source resistor is used. The quiescent output voltage lies between +0.25 and +0.55 V.

Circuit 3-15b is another example of source-resistor biasing with a $-V_{SS}$ supply added. The advantage over circuit 3-15a is that the signal voltage can swing negative to approximately $-V_{SS}$. Two bias lines are shown, one for $V_{SS} = -15$ V and the other for $V_{SS} = -1.6$ V (see Fig.

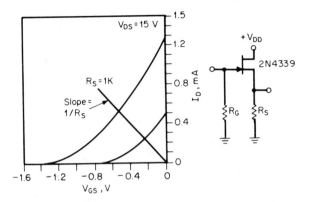

FIGURE 3-16 Self-biasing (Fig. 3-15a) uses the voltage dropped across the source resistor R_S to bias the gate. The load line passes through the origin and has a slope of $-1/R_S$. (Reprinted with permission from J. S. Sherwin, "Build Better Source Followers 10 Ways," *Electronic Design*, Vol. 18, No. 12, copyright Hayden Publishing Co., Inc., 1970.)

3-15). For the first case, the quiescent output voltage lies between $+0.18$ and $+0.74$ V. For the second, it lies between $+0.3$ and $+0.82$ V.

The bias load line for circuit 3-15c is just a horizontal line ($I_D =$ constant). The quiescent output voltage is between $+0.15$ and 0.7 V for $I_D = 0.3$ mA.

Circuit 3-15d is similar to 3-15c except that the $V_{GS} = 0$ output characteristic of FET Q_2 is used as a current source. As seen in Fig. 3-18,

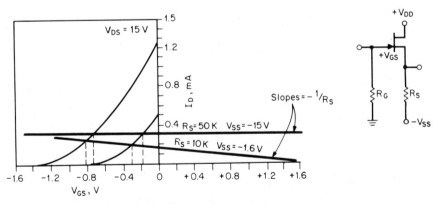

FIGURE 3-17 Adding a V_{SS} supply to the self-bias circuit (Fig. 3-15) allows it to handle large negative signals. The load line's intercept with the V_{GS} axis is at V_{SS}. Bias lines are shown for $V_{SS} = -15$ V and $V_{SS} = -1.6$ V. (Reprinted with permission from J. S. Sherwin, "Build Better Source Followers 10 Ways," *Electronic Design*, Vol. 18, No. 12, copyright Hayden Publishing Co., Inc., 1970.)

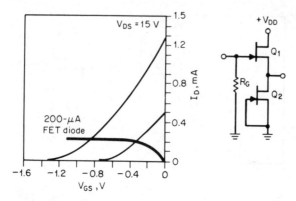

FIGURE 3-18 FET Q_2 does not behave like an ideal current source when its V_{DS} gets very small (Fig. 3-15d). Therefore, Q_1 should have a significantly larger $V_{GS(off)}$ than Q_2. (Reprinted with permission from J. S. Sherwin, "Build Better Source Followers 10 Ways," *Electronic Design*, Vol. 18, No. 12, copyright Hayden Publishing Co., Inc., 1970.)

Q_2 does not supply constant current when its V_{DS} gets very small. This technique should, therefore, be used only to bias FETs whose V_{GS} at the desired I_D is higher than the equivalent $V_{GS(off)}$ of the current-source FET diode.

A pair of matched FETs is used in the circuit of Fig. 3-15e, one as a source follower and the other as a current source. The operating drain current I_{DQ} is set by R_{S2}, as indicated by the load line of Fig. 3-19. The drain current may be anywhere from 0.20 to 0.42 mA, as shown by the limiting transfer characteristic intercepts; however, $V_{GS1} = V_{GS2}$ because the FETs are matched.

Since $I_{D1} = I_{D2}$ and $V_{GS1} = V_{GS2}$, choosing $R_{S1} = R_{S2}$ will ensure that the voltage from point A to B equals the voltage from point C to D (Fig. 3-15e). This source follower, therefore, exhibits zero or near-zero offset. If the FETs are temperature-matched at the operating I_D, the source follower will exhibit zero or near-zero temperature drift.

3-9-2 Biasing with Feedback Increases Z_{in}

Each of the feedback-type source followers (Fig. 3-15f through k) is biased by a method similar to that used with the nonfeedback circuit above it. However, in each case, R_G is returned to a point in the source circuit that provides almost unity positive feedback to the lower end of R_G. If R_S is chosen so that R_G is returned to zero dc volts (except in circuits 3-15f and j), then the input/output offset is zero. R_1 is usually much larger than R_S.

Circuit 3-15f is useful principally for ac-coupled circuits. R_S is usually much less than R_1 to provide near-unity feedback. The bias load line is set by R_S (Fig. 3-20). The output load line, however, is determined by the sum of $R_S + R_1$. The feedback voltage V_{FB}, measured at the junction of R_S and R_1, is determined by the intercept of the $R_S + R_1$ load line with the V_{GS} axis. The quiescent output voltage V_S is $V_{FB} - V_{GS}$.

In the circuit of Fig. 3-15g, R_S can be trimmed to provide zero offset. As the curves of Fig. 3-21 show, R_S will be between 670 Ω and 2.5 kΩ. R_S is much less than R_1. The source load line intercepts the V_{GS} axis at $V_{SS} = -15$ V.

Circuit 3-15h is almost the same as 3-15g; the difference is that resistor R_1 is replaced by a current source. Since an ideal current source has infinite impedance, the bias curve of circuit 3-15h differs from that of Fig. 3-15g (Fig. 3-22) in that the load line is perfectly flat.

Circuit 3-15j is similar to 3-15h except that the output is taken from the top of R_S to reduce the output impedance. R_S must be trimmed if the circuit is to work at all properly.

In Fig. 3-22, the constant-current load line represents a 0.3-mA current source, and the effect of a 1-kΩ source resistor is shown. The offset voltage is seen to lie between 0.2 and 0.75 V. The intercept of

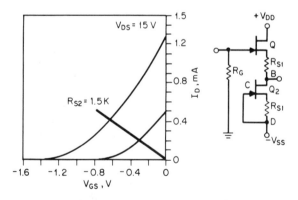

FIGURE 3-19 This load line is set by R_{SS} and Q_2, which acts as a current source (Fig. 3-15e). If its components are properly matched, the circuit will have zero or near-zero offset. (Reprinted with permission from J. S. Sherwin, "Build Better Source Followers 10 Ways," *Electronic Design*, Vol. 18, No. 12, copyright Hayden Publishing Co., Inc., 1970.)

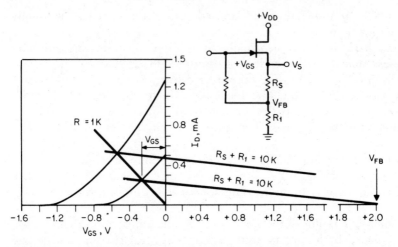

FIGURE 3-20 The bias load line is set by R_S, but the output load line is determined by $R_S + R_1$ when gate feedback is employed (Fig. 3-15*f*). The feedback V_{FB} is determined by the intercept of the $R_S + R_1$ load line and the V_{GS} axis $(V_{GS} \triangleq V_{FB} - V_S)$. (Reprinted with permission from J. S. Sherwin, "Build Better Source Followers 10 Ways," *Electronic Design*, Vol. 18, No. 12, copyright Hayden Publishing Co., Inc., 1970.)

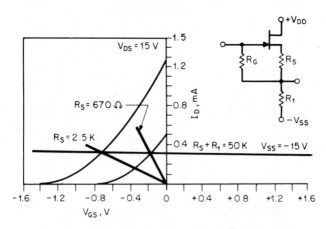

FIGURE 3-21 R_S can be trimmed to provide zero offset at some point between 670 Ω and 2.5 kΩ (Fig. 3-15*g*). The source load line intercepts the V_{GS} axis at $V_{SS} = -V_{GG} = -15$ V. Note that this load line is not perfectly flat. It has a slope of $-1/50$ K, because the current source is not perfect; it has a resistance of 50 kΩ. (Reprinted with permission from J. S. Sherwin, "Build Better Source Followers 10 Ways," *Electronic Design*, Vol. 18, No. 12, copyright Hayden Publishing Co., Inc., 1970.)

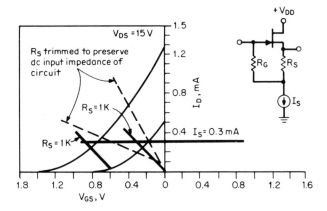

FIGURE 3-22 If R_S is not trimmed so that the load line passes through the origin, a voltage will appear at the gate, causing a reduction in dc input impedance. The incremental impedance will not be affected. (Reprinted with permission from J. S. Sherwin, "Build Better Source Followers 10 Ways," *Electronic Design*, Vol. 18, No. 12, copyright Hayden Publishing Co., Inc., 1970.)

the R_S load line and the V_{GS} axis sets the voltage at the junction of R_S and the current source (V_{FB}). For $R_S = 1$ kΩ, V_{FB} will be between -0.1 and $+0.45$ V. Since V_{FB} appears at the gate, it must be zero if the dc input impedance of the circuit is to be preserved. This can be done by trimming R_s, as shown dashed in Fig. 3-22. The biasing then becomes the same as for circuit 3-15*h*.

Biasing for circuit 3-15*k* is identical to that for circuit 3-15*e* except that feedback is added to raise the input impedance.

Figure 3-23 shows the general source-follower circuit, the equivalent small-signal, low-frequency equivalent circuit, and equations.

3-10   DIFFERENTIAL AMPLIFIER

A "differential amplifier" is one that provides an output proportional to the *difference* between the signals at the two input terminals while preventing an output occurring from a signal which is *common* to the two input terminals. The input terminals are "floating"—that is, neither input is grounded. The signal of interest is impressed between the two inputs. The output may be either a differential floating output or a single-ended output referenced to ground.

Good dc amplifier operation depends upon low-drift design. A single-

100

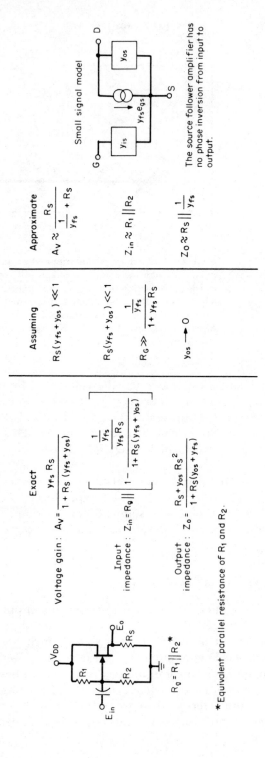

Small signal model

The source follower amplifier has no phase inversion from input to output.

Voltage gain: $A_v = \dfrac{y_{fs} R_S}{1 + R_S(y_{fs} + y_{os})}$

Input impedance: $Z_{in} = R_g \left\| \dfrac{\dfrac{1}{y_{fs}}}{y_{fs} R_S} \left[1 - \dfrac{1}{1 + R_S(y_{fs} + y_{os})} \right] \right.$

Output impedance: $Z_o = \dfrac{R_S + y_{os} R_S^2}{1 + R_S(y_{os} + y_{fs})}$

Exact

Assuming

$R_S(y_{fs} + y_{os}) \ll 1$

$R_S(y_{fs} + y_{os}) \ll 1$

$\dfrac{1}{y_{fs}}{1 + y_{fs} R_S}$

$R_G \gg$

$y_{os} \longrightarrow 0$

Approximate

$A_v \approx \dfrac{R_S}{\dfrac{1}{y_{fs}} + R_S}$

$Z_{in} \approx R_1 \| R_2$

$Z_o \approx R_S \| \dfrac{1}{y_{fs}}$

$R_g = R_1 \| R_2^*$

*Equivalent parallel resistance of R_1 and R_2.

FIGURE 3-23 Summary of equations of source-follower amplifier (low-frequency model).

ended transistor stage may have drift which exceeds the magnitude of the signal to be amplified. Taking advantage of a pair of matched transistors and a design which provides common-mode rejection, the differential amplifier makes possible low-drift dc amplifiers.

3-10-1 Basic Design

Figure 3-24 shows a source-coupled pair of FETs. Assume that Q_1 and Q_2 are matched—that is, that they have identical transfer and output characteristics. With the input, e_i, at zero and e_{cm} at zero, the current source I_S will divide equally between Q_1 and Q_2, so that $I_{D1} = I_{D2}$ and $V_{D1} = V_{D2}$. The differential output voltage e_o will be zero. A change in the value of I_S or V_{DD} will not cause a differential output voltage. Now assume an input voltage e_i of +50 mV. This positive voltage tends to increase the current through Q_1. Any increase in I_{D1} must be offset by a corresponding *decrease* in I_{D2}, since the total current is fixed at I_S. If V_S were fixed, then ΔI_{D1} would be $e_i g_{fs}$; however, V_S is not fixed. Since ΔI_{D2} must equal minus ΔI_{D1}, v_{gs2} must equal minus v_{gs1} (assuming that g_{fs} of the two FETs is constant for small signals). Therefore, we can assume that

$$v_{gs1} = -v_{gs2} = \tfrac{1}{2}e_i \tag{3-27}$$

(This assumes that $g_{fs1} = g_{fs2}$.)

The voltage changes at D_1 and at D_2 are

$$-v_{d1} = v_{gs1}g_{fs}R_D = \tfrac{1}{2}e_i g_{fs}R_D \tag{3-28}$$

$$v_{d2} = v_{gs2}g_{fs}R_D = -\tfrac{1}{2}e_i g_{fs}R_D \tag{3-29}$$

therefore

$$e_o = v_{d1} - v_{d2} = -e_i g_{fs}R_D \tag{3-30}$$

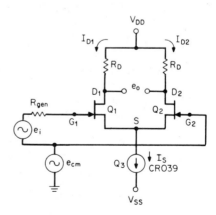

FIGURE 3-24 Differential amplifier (dual JFET type U401).

Assume $g_{fs} = 10^{-3}$ mho and $R_D = 10$ kΩ. The output voltage with a 50-mV input signal is

$$e_o = -0.05 \times 10^{-3} \times 10,000 = -0.5 \text{ V}$$

The voltage amplification of this differential stage is

$$A_V = \frac{e_o}{e_i} = -g_{fs} R_D \qquad (3\text{-}31)$$

It is the same as a single-stage common-source amplifier. [Output conductance g_{os} is neglected in Eqs. (3-28) through (3-31).]

A significant advantage of the differential amplifier is its ability to differentiate between common-mode input and differential input signals. Consider an input signal to be amplified e_i and an undesired signal e_{cm} to be rejected. The differential signal will result in an output voltage e_o given by Eq. (3-30). The common-mode signal, however, is applied to both gates equally, and if the devices are matched, there will be no differential change in drain currents, and hence there will be no differential output e_o.

The common source S of the two FETs will follow the common-mode voltage (offset by $-V_{GS}$); therefore V_{DS} will change by the magnitude of the common-mode voltage. V_{DS} should not be permitted to fall below the "pinchoff" region of the FET output characteristics. This condition will be met if $V_{DG} \geq -V_{GS(off)}$; therefore, in the positive direction, the common-mode voltage should not exceed

$$\begin{aligned} e_{cm(max)} &= V_D + V_{GS(off)} \\ &= V_{DD} - I_D R_D + V_{GS(off)} \end{aligned} \qquad (3\text{-}32)$$

(Note: $V_{GS(off)}$ for the n-channel FET is a negative number.)

In the negative direction e_{cm} should not cause V_S to drop below the value required for the current source Q_3 to maintain its function of supplying adequate source current I_S. Other limits on e_{cm} are imposed by the maximum V_{DG} rating of the FETs and by the V_{DG} that causes I_G to become excessive.

The voltage-gain equation (3-31) shows that A_V is a linear function of R_D. For maximum gain a high value of R_D is desirable. Given a supply voltage, increasing the value of R_D decreases V_D, unless I_S is decreased to keep $R_D I_D$ constant. The choice of R_D is affected by:

1. Desired voltage gain A_V

2. Available supply voltage V_{DD}

3. Maximum common-mode input voltage $e_{cm(max)}$

4. Maximum value of common-source current $I_{S(max)}$

5. Maximum differential input voltage e_i

6. High-frequency roll-off

We will go through some of the design steps for the circuit shown in Fig. 3-24. The following assumptions are made:

1. Common-mode voltage: $e_{cm} = \pm 8$ V.

2. Differential input voltage: $e_i = \pm 0.05$ V.

3. Voltage amplification: $A_V =$ as high as practical.

4. Available supply voltages: $V_{DD} = 15$ V $\pm 5\%$; $V_{SS} = -15$ V $\pm 5\%$. (Factors affecting the choice of the proper FET pair for Q_1 and Q_2 are discussed in Sec. 3-6. We will assume for this example that a general-purpose dual type has been selected.)

5. Q_1 plus Q_2 — dual FET type U401. (The matched characteristics of the U401 are specified at an operating I_D of 200 μA; therefore, we will select a current source Q_3 of approximately twice this value.)

6. Q_3 — Current regulator diode type CR039.

The maximum value of R_D is arrived at by solving for R_D in the equation

$$V_{DD(min)} - I_{D(max)} R_D = V_{D(min)} - V_{CM(max)} + V_{i(max)} + V_{DG(min)} \quad (3\text{-}33)$$

As a rule of thumb, to stay in the saturated current region of the FET output characteristics, V_{DG} should not fall below the value of $- V_{GS(off)(max)} = 2.5$ V; therefore, we set $V_{DG(min)} = 2.5$ V. $I_{D(max)}$ will occur when the differential input is at $e_{i(max)}$ and is given by

$$I_{D(max)} = \tfrac{1}{2} I_{S(max)} + \tfrac{1}{2} e_{i(max)} g_{fs(max)} \quad (3\text{-}34)$$

From the data sheets for the CR039 and the U401, $I_{S(max)} = 1.1 \times 0.390$ mA and $g_{fs(max)} = 1.6$ mmhos. Solving Eq. (3-33) for R_D yields

$$R_{D(max)} = \frac{V_{DD(min)} - V_{CM(max)} - e_{i(max)} - V_{DG(min)}}{\tfrac{1}{2}(I_{S(max)} + e_{i(max)} g_{fs(max)})}$$

$$R_{D(max)} = \frac{2(14.25 - 8 - 0.05 - 2.5)}{(1.1)(0.390)(10^{-3}) + (0.05)(1.6)(10^{-3})} \quad (3\text{-}35)$$

$$R_{D(max)} = 14{,}538 \ \Omega$$

We will use a value of 14,000 Ω for R_D and calculate the voltage gain using the data-sheet *minimum* value for g_{fs}.

$$A_V = -g_{fs}R_D = -(1,000)(10^{-6}) \times 14,000 = -14$$

We will now consider ways in which A_V may be increased, while still adhering to assumptions 1 through 4 above. Equation (3-31) shows that to increase A_V, the $g_{fs}R_D$ product must be increased. Other factors remaining constant, if R_D is increased, then $I_{D(max)}$ must be decreased to maintain the same minimum value for V_D. A decrease in I_D will result in a decrease in g_{fs}; however, as shown in Eq. (2-6) and Fig. 2-19, for a given FET the g_{fs} decrease will be proportional to the square root of the I_D decrease. Operating at a lower value of I_D, with a given device, *can* result in a higher voltage gain. In this example, if we reduce the value of I_S by a factor of 2, g_{fs} will be reduced by a factor of only $2^{1/2}$. Doubling R_D, then (to maintain the same $I_D R_D$), will increase A_V by a factor of $2^{1/2}$; thus the voltage gain would be increased from 14 to 19.8. Because of the doubling of R_D, the output resistance would be increased and the high-frequency corner would be decreased. Also, if the FETs are not operated at the data-sheet-specified value of I_D, the offset and drift specifications may not be met. Increasing the value of V_{DD} would also permit a higher value of R_D while maintaining the same $I_{D(max)}$. Care must be taken not to exceed the $V_{DG(max)}$ voltage rating of the FET.

3-10-2   Drift and Offset Compensation

JFET differential inputs are often used in dc amplifiers because they provide high input impedance and operate with extremely low input bias current. To achieve accurate measurements, the input offset and equivalent input drift must be made small with respect to the signal being measured. Methods of nulling or minimizing input offset and drift and how this compensation affects the common-mode rejection ratio (CMRR) will be examined.

An understanding of the effect of temperature changes upon the basic parameters of the FET aids in the application of temperature-induced drift compensation. For the purpose of this discussion, temperature drift is defined as the change in V_{GS} required to maintain a constant I_D as the temperature is changed. For a typical JFET, operating with a V_{DG} greater than $-V_p$, the relationship between I_D and V_{GS} was shown in Eq. (3-4).

As pointed out in Chapter 2, V_p is a function of the zero-bias depletion-layer width at the gate-channel p-n junction. The depletion width decreases with increasing temperature, which results in an increase in channel width and therefore an increase in $-V_p$. The temperature coefficient of $-V_p$ is about 2.2 mV/°C. This widening of the channel with increasing temperature tends to increase channel conduction; *however*, at the same time, increasing temperature causes a *decrease* in carrier mobility and thus in channel conductivity. Carrier mobility has a negative temperature coefficient of about 0.6%/°C. The positive channel-thickness coefficient and the negative mobility coefficient have opposite effects upon the temperature characteristic of I_D and I_{DSS}. In a thick-channel device (high $V_{GS(off)}$) the mobility factor dominates and I_{DSS} has a negative temperature coefficient. In a thin-channel unit the channel-thickness effect dominates and I_{DSS} has a positive temperature coefficient. At a particular channel thickness the temperature coefficients exactly compensate and I_{DSS} has a zero temperature coefficient. It has been shown that this particular thickness occurs in JFET devices having $V_{GS(off)}$ values of about -0.63 V. Units having *higher* $V_{GS(off)}$ values can be biased down to this critical channel dimension, so that I_D has zero temperature drift. Mathematically the zero-drift drain current I_{DZ} is

$$I_{DZ} = I_{DSS}\left(\frac{0.63}{V_{GS(off)}}\right)^2 \tag{3-36}$$

Operating at drain currents other than I_{DZ} will result in a drift in V_{GS} as given by the equation

$$V_{GS(drift)} \approx 2.2 \text{ mV/°C}\left[1 - \left(\frac{I_D}{I_{DZ}}\right)^{1/2}\right] \tag{3-37}$$

A plot of this characteristic is shown in Fig. 3-25. Differentiating Eq. (3-37) with respect to I_D produces a value K for the amount of drift change per unit change of I_D:

$$\frac{dV_{GS(drift)}}{dI_D} = \frac{(-1)(10^{-3})}{(I_D/I_{DZ})^{1/2}} = K \quad \text{V/°C} \tag{3-38}$$

FIGURE 3-25 Drift of gate-source voltage in a single JFET drops with increasing normalized drain current. The slope of this curve is useful in determining the size of differential source resistance needed to null a given drift.

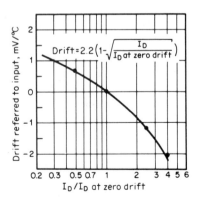

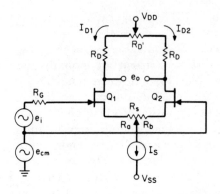

FIGURE 3-26 Differential amplifier with offset and drift compensation.

Ideally the differential amplifier produces an output only as a function of a differential input signal. If the temperature-induced drifts (dV_{GS}/dT) of the two FETs are equal, they appear to be a common-mode signal and do not produce an output. Unfortunately, few dual FETs have a perfect V_{GS} drift match; thus a differential offset will occur.

Since V_{GS} drift depends upon the value of drain current, the differential drift d may be reduced or eliminated by unbalancing the two drain currents by an amount ΔI_D:

$$\Delta I_D = \frac{d}{K} \qquad (3\text{-}39)$$

where K is given by Eq. (3-38).

With the input voltage e_i held to zero, drain-current unbalance can be induced by adding unbalanced source resistors as shown in Fig. 3-28. A potentiometer R_S is used so that the unbalance can be adjusted. The resistor unbalance is a function of g_{fs}, I_S, and the required current unbalance.

$$\Delta I_D = I_{D2} = I_{D1} = \tfrac{1}{2}[g_{fs}I_S(R_a - R_b)] \qquad (3\text{-}40)$$

Substituting Eq. (3-39) for ΔI_D and solving for R, the value of the R_S necessary to null a given drift becomes

$$R_S = R_a - R_b = \frac{2d}{g_{fs}I_S K} \qquad (3\text{-}41)$$

Minimum values of g_{fs}, I, and I_S are used to ensure an adequate value for R_S.

Even if the FETs had no output initially, the introduction of the source resistance R_S would probably introduce an offset. This offset is due to the unbalanced current ΔI_D flowing through the balanced load resistors R_D. In this case, output balance can be achieved by unbalancing the load resistors R_D to compensate for the unbalance in I_D. The potentiometer R_D in the drain circuit of Fig. 3-26 serves this function. Since the output offset will be A_V times the input offset, the drain-offset compensation resistor will need to be A_V times the source-drift-compensating resistor R_S.

$$R_{D1} = A_V R_S = g_{fs} R_D R_S \qquad (3\text{-}42)$$

The actual value of R_D' should be greater than this so that any initial FET pair offset can be compensated for. The value that takes this into account is

$$R_D' = g_{fs} R_D \left[R_S + \frac{2(V_{GS1} - V_{GS2})}{I_S} \right] \qquad (3\text{-}43)$$

where $V_{GS1} - V_{GS2}$ = maximum specified offset of the FET pair. With the circuit shown in Fig. 3-26, first R_S would be adjusted for minimum drift with temperature, then R_D' would be adjusted for zero offset.

If the differential FET pair is used as the input stage of an operational amplifier (op amp), as shown in Fig. 3-27, the drift-adjusting procedure is modified. This occurs because with the feedback loop closed, the high-gain amplifier tends to force $V_{D1} - V_{D2}$ to zero. Thus to introduce an unbalance in drain current (ΔI_D), the drain potentiometer is first adjusted. This causes an output offset which is then eliminated by adjusting the source potentiometer R_S. Assuming the amplifier offset is zero, the calculation of the value of R_D' and R_S' is the same as for the single-stage amplifier. Compensation of the op amp offset and drift can be achieved by modifying the calculations for R_{D1} and R_S. The amplifier drift divided by the input stage gain ($g_{fs} R_D$) should be added to the value d in Eq. (3-39).

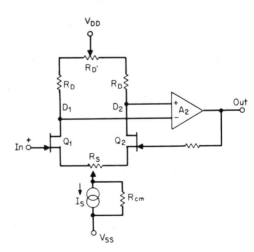

FIGURE 3-27 Differential JFET input for operational amplifier.

The amplifier offset divided by the input stage gain should be added to the value $V_{GS1} - V_{GS2}$ in Eq. (3-43). Typically, this will change R'_D and R'_S by less than 10 percent.

Additional A_2 offset and drift will result from A_2 input-current offset and drift times the value of $R'_D + 2R_D$. Again, this is referred to the input by dividing by the voltage gain (A_V) of the input stage.

3-10-3   Common-Mode Errors

In the ideal differential amplifier a common-mode voltage e_{cm} would cause no differential output voltage e_o. The amplifier output would be a function only of the differential input e_i and would reject any common-mode input signal. In practice, this does not occur because nonideal conditions exist. One important figure of merit is a measure of the degree to which the differential amplifier rejects the common-mode signal in favor of the differential signal. This figure of merit is called the common-mode rejection ratio (CMRR).

Common-mode input signals cause two types of output errors in differential amplifiers. One type of error is a differential output signal caused by the common-mode input; the other is a common-mode output signal where both outputs change equally and in the same direction. Since the op amp which follows the FET input stage will typically amplify the differential-mode error 80 dB more than the common-mode error (assuming its CMRR is 80 dB), the differential-mode error is by far the more important of the two.

Differential-mode error is the result of unequal gain between the two sides of the differential amplifier. This can be caused by imprecise transconductance matching of the dual FETs, unbalanced drain resistances, unbalanced source resistances, imprecise output conductance matching of the FETs, or any combination of these factors. Thus, if the drain or source resistances are adjusted for drift or offset nulling, the differential output error will be increased and CMRR degraded.

Consider the case where the differential amplifier has slightly unbalanced gain. An equivalent circuit is shown in Fig. 3-28.

A figure of merit CMRR can be derived for differential output errors:

$$\text{CMRR}_{(\text{diff})} = \frac{A_{\text{diff}}}{A_{\text{CM(diff)}}} \tag{3-44}$$

The differential-mode gain, assuming both halves of the differential amplifier are nearly equal, can be approximated as

$$A_{\text{diff}} \approx \frac{g_{fs}R_D}{1 + g_{fs}R_S}\left(\frac{1}{1 + g_{os}R_D}\right) \tag{3-45}$$

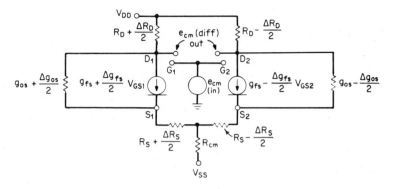

FIGURE 3-28 Differential output equivalent circuit.

for $1/g_{os} \gg R_S$. For $R_S \ll 1/g_{fs}$ and $R_D \ll 1/g_{os}$, this simplifies to the familiar expression

$$A_{\text{diff}} \approx g_{fs} R_D \tag{3-46}$$

If the circuit is balanced except for the drain resistors, the differential-mode error resulting from a common-mode input signal becomes

$$A_{cm(\text{diff})} \approx \frac{\Delta R_D}{2 R_{cm}} \tag{3-47}$$

so

$$\text{CMRR}_{(dB)} = 20 \log \left(\frac{2 R_{cm}}{\Delta R_D} g_{fs} R_D \right) \tag{3-48}$$

Similarly, the error introduced by unbalanced transconductances is taken into account:

$$\text{CMRR}_{(dB)} = 20 \log 2 \, \Delta g_{fs} R_{cm} \tag{3-49}$$

When unequal output conductances are taken into account,

$$\text{CMRR}_{(dB)} = 20 \log \frac{2 g_{fs} R_{cm} R_o}{\Delta R_o} \tag{3-50}$$

and with unbalanced source resistors,

$$\text{CMRR}_{(dB)} = 20 \log \frac{2 R_{cm}}{\Delta R_S} \tag{3-51}$$

assuming $1/g_{fs} \gg R_S$.

A more general equation, resulting from the combination of all the above factors, is

$$\text{CMRR}_{(dB)} = 20 \log$$

$$\left[\frac{2g_{fs}R_{cm}}{(\Delta R_S + 1/\Delta g_{fs})/(R_S + 1/g_{fs}) \pm \Delta R_D/R_D \pm \Delta R_o/R_o} \right] \quad (3\text{-}52)$$

The other type of error, common-mode output voltage, is important when only one of the outputs of the differential stage is used, or when the following stage has a poor CMRR. Figure 3-29 shows how common-mode output error can be measured. If common-mode gain for common-mode output errors is defined as

$$A_{CM(cm)} = \frac{V_{D1}}{V_{G1}} = \frac{V_{D2}}{V_{G2}} \quad (3\text{-}53)$$

then an expression may be found by noting that

$$V_D = -g_{fs}V_{GS}R_D \quad (3\text{-}54)$$

However,

$$V_{GS} = V_G - V_S \quad (3\text{-}55)$$

and

$$V_S = \frac{-2V_D R_{CM}}{R_D} \quad (3\text{-}56)$$

Thus,

$$V_D = -g_{fs}V_G R_D - g_{fs}2R_{CM}V_D \quad (3\text{-}57)$$

and

$$A_{CM(cm)} = \frac{V_D}{V_G} = \frac{-g_{fs}R_D}{1 + 2g_{fs}R_{CM}} \quad (3\text{-}58)$$

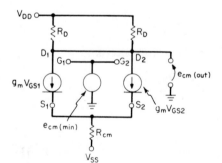

FIGURE 3-29 Common-mode output equivalent circuit.

It is common practice to assign a figure of merit to a differential amplifier. This figure of merit is "common-mode rejection ratio," usually abbreviated as $\text{CMRR}_{(cm)}$. $\text{CMRR}_{(cm)}$ is defined as

$$CMRR_{(cm)} = \frac{A_{\text{diff}}}{A_{CM(cm)}} \qquad (3\text{-}59)$$

When Eqs. (3-31) and (3-58) are substituted into Eq. (3-59), it is apparent that the common-mode rejection ratio for common-mode output errors is

$$CMRR_{(cm)} \approx \frac{-g_{fs}R_D}{-g_{fs}R_D/(1 + 2g_{fs}R_{CM})} = 1 + 2g_{fs}R_{CM} \qquad (3\text{-}60)$$

Most op amps have single-ended outputs. In such cases, there can be only one result of a common-mode input signal: an error in the output. In a FET input op amp the total CMRR will be a function of the common-mode and differential output errors of the preamplifier as well as the CMRR of the second stage. To calculate the combined CMRR of both stages of the FET input op amp, refer to Eq. (3-59), which is the definition of CMRR.

A two-stage amplifier is shown in Fig. 3-30. CMRR in this circuit is calculated as follows:

$$CMRR = \frac{A_{\text{diff}}}{A_{CM}} \qquad (3\text{-}61)$$

$$= \frac{(A_{\text{diff1}})\,(A_{\text{diff2}})}{(A_{CM(cm)})\,(A_{CM2}) + (A_{CM(\text{diff})1})\,(A_{\text{diff2}})}$$

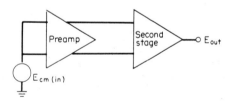

FIGURE 3-30 Two-stage op amp CMRR model.

Substituting Eq. (3-59) into Eq. (3-61), we have

$$CMRR_{\text{total}} = \frac{1}{1/(CMRR_{(cm)1})\,(CMRR_2) + 1/(CMRR_{(\text{diff})1}} \qquad (3\text{-}62)$$

where $CMRR_{(cm)1}$ is the CMRR referred to in Eq. (3-60) for the FET preamplifier, $CMRR_2$ is the CMRR specification of the second stage, and $CMRR_{(\text{diff})1}$ is the CMRR referred to in Eq. (3-52) for the FET preamplifier.

In a common-drain stage, the common-mode gain with respect to common-mode output error ($A_{\text{CM}(cm)}$) is unity (when $A_{\text{diff}} = 1$). Hence, the entire output common-mode signal is passed along to the second stage. All the equations in this section, except for Eqs. (3-53), (3-58), and (3-60), apply equally to common-drain preamplifiers.

3-10-4 Frequency Response

The frequency response of a FET differential amplifier is determined by two time constants, one for the input and one for the output.

The input time constant is formed by the generator impedance and the effective input capacitance C_{in}, where

$$C_{\text{in}} = \left(1 + \frac{g_{fs}R_D}{1 + g_{fs}R_S}\right)C_{DG} + \left(1 - \frac{g_{fs}R_S}{1 + g_{fs}R_S}\right)C_{GS} + C_{\text{stray}} \qquad (3\text{-}63)$$

and

$$f_{\text{in}} = \frac{1}{2\pi C_{\text{in}} R_G} \qquad (3\text{-}64)$$

The output time constant is formed by the drain resistance and the output capacitance C_{out}, where

$$f_{\text{out}} = \frac{1}{2\pi C_{\text{out}} R_D} \qquad (3\text{-}65)$$

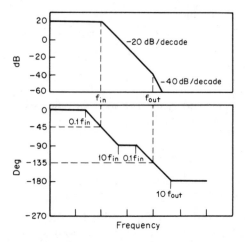

FIGURE 3-31 Phase and gain Bode plots for typical JFET preamplifier.

and

$$C_{out} = C_{gd} + C_{load} + C_{stray} \qquad (3\text{-}66)$$

Equations (3-63) and (3-64) apply to common-drain preamplifiers as well as to common-source preamplifiers if $R_D = 0$.

For the case of the output time constant for the common-drain preamplifier,

$$f_{out} = \frac{g_{fs}}{2\pi C_{out}} \qquad (3\text{-}67)$$

where

$$C_{out} = C_{gs} + C_{sd} + C_{load} + C_{stray} \qquad (3\text{-}68)$$

Typically, the Bode plot of a FET preamplifier will resemble that shown in Fig. 3-31.

3-10-5 Stability and Phase Compensation

The stability criterion for any closed-loop system, including an operational amplifier, is that the phase shift around the loop must never reach 360° for any frequency at which the gain is unity or greater. Since the feedback around an op amp is in the inverting input, there is an intrinsic 180° phase shift. This dictates that the additional phase shift in the amplifier plus the phase shift of the feedback network must be less than 180° for all frequencies where gain is unity or greater.

Most commercially available operational amplifiers are compensated so that they have only 90° of phase shift where gain is unity or greater. This is demonstrated in Fig. 3-32. When a FET preamplifier is added ahead of an op amp, the total gain and phase response will be the sum of the response of the preamplifier and of the second stage. Often, this results in a phase shift of 180° or greater over much of the frequency range. An intolerable situation is thus created in that the total FET-input op amp will oscillate if the gain is set to a value where the phase shift is greater than 180°.

As a rule of thumb, an op amp will not oscillate if the closed-loop gain is such that the slope of the rolloff is no greater than 20 dB/decade. This point corresponds to a worst-case phase shift of 135°.

3-10-6 Selecting the Proper FET Pair

FET differential pairs can be broken down into four general types, according to their intended application. These types include low-leakage, low-noise, high-frequency, and general-purpose FET duals.

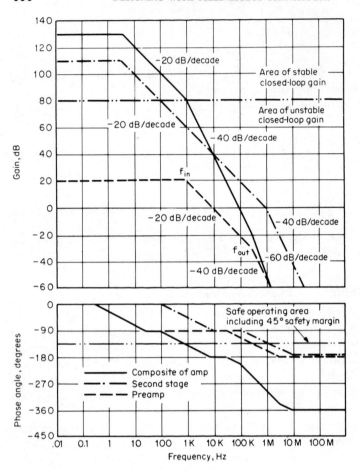

FIGURE 3-32   Compensated phase shift in op amps.

Low-leakage FETs generally have leakage currents in the range of 0.1 to 1.0 pA at 25°C. To achieve this low leakage, the active area of the device is made small. This small active area produces low g_{fs} and low capacitance. Although low-leakage FETs are preferred whenever low circuit leakage is the primary design criterion, the designer should consider using a general-purpose FET if slightly higher leakage can be tolerated. General-purpose devices offer better g_{fs}, offset, drift, and $\bar{e}_n$ than do low-leakage devices. One feature of the low-leakage FET which is not often specified in data sheets, but which is important to actual circuit performance, is the I_G breakpoint, that is, the drain-to-gate voltage at which the gate current rises rapidly. In practical circuits, the drain-to-gate voltage can reach 15 or 20 V, causing excessive leakage for FETs with low I_G breakpoints. If a common-mode voltage range in excess of 10 to 15 V is required, a Cascode input stage should be considered.

Low-noise FETs are designed primarily for low $\bar{e}_n$. They also have moderate g_{fs}, low g_{os}, and moderate leakage and breakdown voltage. Low-noise devices tend to have better drift and CMRR characteristics than do other types of dual FETs. A low-noise FET can produce the lowest noise operation of any FET when the generator impedance is below 1 to 10 MΩ. For higher generator impedances, low-leakage FETs may well provide the lowest overall noise performance of any transistor, bipolar or JFET, because of their lower noise current.

High-frequency dual FETs have very high g_{fs} and low capacitance. To achieve these characteristic leakage currents, breakdown voltage and the I_G breakpoint must be sacrificed. Because of these performance trade-offs, high-frequency FETs should be used only when their high-gain bandwidth is a design requirement. Dual high-frequency FETs are usually of hybrid (two-chip) construction rather than of a monolithic design because of the significantly lower capacitances between the two chips.

General-purpose dual devices are often the best choice for FET input op-amp applications, since they exhibit good g_{fs} and breakdown voltage and moderately low g_{os}, leakage current, and capacitance. General-purpose dual FETs also tend to have low $\bar{e}_n$ and good CMRR and drift characteristics.

Representative geometries of the four types of FETs just described are shown in Fig. 3-33 and provide a good insight into the relationship of device active area and performance characteristics.

When selecting any dual FET, two factors which affect the overall performance of the devices should be considered. One is the maximum value of $V_{GS(off)}$; a low maximum value of $V_{GS(off)}$ simplifies bias design and improves common-mode range, CMRR, offset, and drift. The other factor is the offset of the device. Although any value of offset can be nulled out of a circuit, the nulling process itself degrades CMRR and the drift performance of the circuit.

3-10-7 Selecting the Op-Amp Integrated Circuit

The monolithic IC op-amp portion of a FET input op-amp circuit either contributes to or solely determines five parameters of the complete circuit. For three of the parameters—offset, drift, and noise—the contributions are diminished in the complete op-amp circuit by the gain of the FET differential amplifier. In many circuits, however, these parameters are still significant and should not be ignored. Whenever possible, an IC op-amp device should be chosen which specifies maximum values

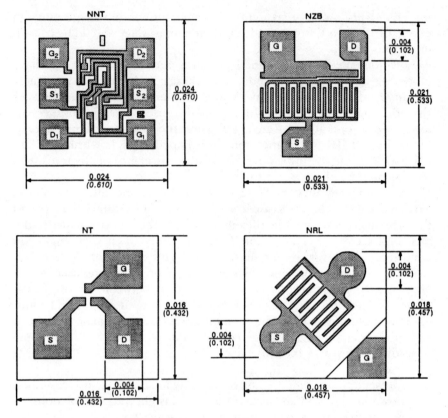

FIGURE 3-33 Four JFET geometries. [All dimensions given in inches (millimeters).] (Courtesy of Siliconix Incorporated.)

for V_{os}, I_{os}, V_{drift}, I_{drift}, $\bar{e}_n$, and $\bar{i}_n$. Many IC op-amp data sheets will provide only typical values for these parameters. Typical values can be in error by an order of magnitude, and almost inevitably vary from lot to lot.

One parameter which is usually ignored on IC op-amp data sheets is so-called popcorn noise. Even though this parameter is missing from the data sheet, it is usually present in the op-amp IC and can be quite troublesome in low-noise designs. If low noise is a prime design criterion, an op amp specified for low popcorn noise should be selected.

Two other circuit parameters, output impedance and slew rate, are determined almost entirely by the IC op-amp portion of the circuit.

3-10-8   Design Example

This section presents a typical FET input op-amp design example (Fig. 3-34) using a Siliconix U401 n-channel junction FET. The U401 is designed to be operated with an I_D of 200 μA per device, or with an I_{CM} of 400 μA. A general-purpose operational amplifier should have a large common-mode range and good CMRR. These two requirements dictate that an active current source be used, such as the Siliconix CR043 current-regulator diode. The CR043 is ideal for this purpose, since it has a nominal current value of 430 μA $\pm$ 10% and a low temperature coefficient of less than 0.05%/°C typically, or 0.2 μA/°C.

Selection of drain resistor value involves a tradeoff between preamplifier gain and common-mode range. Common-mode range decreases and gain increases proportionally to increased value of the drain resistor. If a voltage drop of 3 V across the drain resistors is permissible, then the value of the drain resistors can be found by applying Ohm's law, as in

$$R_D = \frac{3 \text{ V}}{200 \text{ } \mu\text{A}} = 15 \text{ k}\Omega$$

Gain of the preamplifier will be

$$A_{\text{diff}} \approx -g_{fs} R_D \tag{3-69}$$

From the U401 data sheet, g_{fs} at 200 μA operating current varies from a minimum of 1000 μmhos to a maximum of 1600 μmhos. For a worst-case design, the minimum value should be used:

$$A_{\text{diff(min)}} \approx (1 \times 10^{-3}) \cdot (1.5 \times 10^4) = 15.0$$

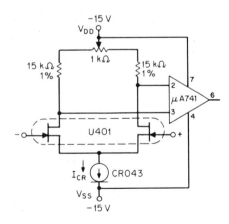

FIGURE 3-34   General-purpose design example circuit.

To keep the FETs in the current-saturated region of their output characteristics, V_{DG} should not fall below $-V_{GS(off)}$. The maximum value of V_G then will be

$$V_{G(max)} = V_{DD} - I_D R_D + V_{GS(off)} \qquad (3\text{-}70)$$

Two factors must be considered in determining the minimum value for V_G. One is the maximum value of V_{DG} determined by either the I_G breakpoint or the maximum V_{DG} rating. The other factor is the minimum voltage needed across the CR043 to maintain it in a high-impedance mode.

If $V_{DG(max)}$ is the limiting factor, then

$$V_{G(min)1} = V_{DD} - I_{D(min)} R_D - V_{DG(max)} \qquad (3\text{-}71)$$

If the minimum voltage $V_{CR(min)}$ across CR043 is the limiting factor, then

$$V_{G(min)2} = V_{SS} + V_{CR(min)} + V_{GS(min)} \qquad (3\text{-}72)$$

Both $V_{G(min)1}$ and $V_{G(min)2}$ should be computed and the most positive value used in the design. For our example the U401 data sheet $V_{DG} = 15$ V for the I_G specification, so that will be the value used for $V_{DG(max)}$. The CR043 data sheet indicates that the minimum "knee impedance" is measured at a V_F of 6 V, so that will be the value used for $V_{CR(min)}$.

The value of $V_{GS(min)}$ can be estimated by substituting the appropriate minimum and maximum values in the general FET V_{GS} equation:

$$V_{GS(min)} = V_{GS(off)(min)} \left[1 - \left(\frac{I_{D(max)}}{I_{DSS(min)}} \right)^{1/2} \right] \qquad (3\text{-}73)$$

From the U401 data sheet, $V_{GS(off)(min)}$ and $I_{DSS(min)}$ are -0.5 V and 0.5 mA. Maximum I_F of the CR043 is 473 μA; therefore $I_{D(max)}$ is $(\frac{1}{2})$ (473) $= 236.5$ μA. Using these values in Eq. (3-73), we get

$$V_{GS(min)} = (-0.5) \left[1 - \left(\frac{0.2365}{0.5} \right)^{1/2} \right] = -0.156 \text{ V}$$

Then with Eqs. (3-70), (3-71), and (3-72) we get

$$V_{G(max)} = 15 - (0.2365)(15) + (-2.5) = 8.95 \text{ V}$$

Minimum I_F of the CR043 $= 0.387$ mA; therefore $I_{D(min)} = 0.1935$ mA.

$$V_{G(min)1} = 15 - (0.1935)(15) - 15 = -2.9 \text{ V}$$

$$V_{G(min)2} = -15 + 6 + (-0.156) = -9.156 \text{ V}$$

If the common-mode input drops below -2.9 V, I_G may become excessive because V_{GD} will exceed 15 V. The common-mode voltage range for the amplifier then is from -2.9 to $+8.95$ V, if the I_G specification is not to be exceeded. If a higher I_G can be tolerated, then the common-mode range can go to -9.156 V.

The offset will be

$$E_{Tin} \approx E_{os1} + \frac{2I_{os}}{g_{fs}} + \frac{E_{os2}}{g_{fs}R_D} \tag{3-74}$$

From the U401 data sheet, $E_{os1} = 5$ mV, and $g_{fs(min)}$ at $I_D = 200$ µA is 1000 µmhos. From the µA741 op-amp data sheet, $I_{os(max)}$ is given as 200 µA. Therefore

$$E_{Tin(max)} = 5 \times 10^{-3} + \frac{2(2 \times 10^{-7})}{1 \times 10^{-3}} + \frac{5 \times 10^{-3}}{15.0} \approx 5.73 \text{ mV}$$

If both sides of Eq. (3-74) are divided by ΔT, then

$$\frac{E_{Tin}}{\Delta T} \approx 10 \times 10^{-6} + \frac{2(6 \times 10^{-10})}{9 \times 10^{-4}} + \frac{1 \times 10^{-5}}{15.0}$$

or

$$\frac{\Delta E_{Tin}}{\Delta T} \approx 12 \text{ µV/°C} \qquad \text{worse case, assuming that the drain resistors are perfectly balanced}$$

For drain nulling, the value of the null potentiometer should be

$$R_N = \frac{2E_{Tin(max)}g_{fs(max)}R_D}{I_{CM(min)}} \tag{3-75}$$

or

$$R_N = \frac{2(5.8 \times 10^{-3})(1.6 \times 10^{-3})(1.5 \times 10^4)}{3.87 \times 10^{-4}}$$

or

$$R_N = 720 \text{ }\Omega$$

However, to this approximation for R_N must be added the maximum drain resistor unbalance, which is 2 percent R_D for 1 percent tolerance resistors, or 300 Ω. Thus $R_N \approx 1$ kΩ. The output frequency of the preamplifier will be

$$f_{out} = \frac{1}{2\pi R_D C_{out}} \tag{3-76}$$

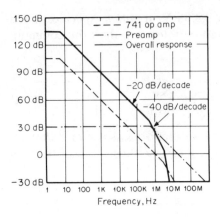

FIGURE 3-35   Bode plot of preamplifier and second stage.

where

$$C_{out} = C_{dg} + C_{load} + C_{stray} \qquad (3\text{-}77)$$

or

$$C_{out} \approx 5 \text{ pF} + 2 \text{ pF} + 5 \text{ pF} \approx 12 \text{ pF}$$

and thus

$$f_{out} = \frac{1}{(6.28)(1.5 \times 10^4)(1.2 \times 10^{-11})} = 880 \text{ kHz}$$

The combined Bode plot of the preamplifier and the second stage is shown in Fig. 3-35. The Bode plot indicates that the op amp will be stable for any closed-loop gain of greater than 35 dB. For closed-loop gains of less than this value, one of the numerous forms of op-amp compensation must be used.

3-10-9   Cascode Differential Amplifiers

The common-mode voltage range of a FET differential amplifier can be increased by utilizing the Cascode configuration. Consider the differential stage shown in Fig. 3-36. The dual FET type U421 was chosen because of its low gate operating current I_G, which is specified as 0.1 pA maximum at $V_{DG} = 10$ V and $I_D = 30$ μA. An examination of the typical performance curves for the U421 reveals that the I_G breakpoint for this device is about 10 V V_{DG}, as shown in Fig. 3-37. This means that even with ideal bias conditions the common-mode voltage range cannot exceed 10 V without an appreciable increase in I_G occurring. Actually the maximum specified value of $-V_{GS\,(off)}$ must be subtracted from the 10 V, leaving a common-mode range of 8 V. A further reduction

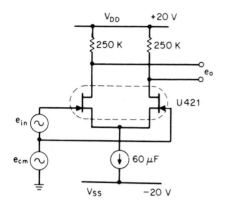

FIGURE 3-36 FET differential amplifier.

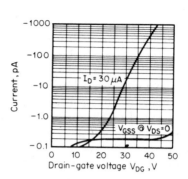

FIGURE 3-37 Leakage currents vs. drain voltage.

must be allowed for $(1 - A_V)/2$ times the maximum differential signal. Thus, even though the U421 has a maximum gate-drain and gate-source voltage rating of -40 V, operating gate current will start to increase appreciably when V_{DG} exceeds about 10 V.

Common-mode voltage range, common-mode input resistance, and common-mode rejection ratio can be improved by the addition of another pair of FETs connected Cascode, as shown in Fig. 3-38. In the example, we utilize the Siliconix type U401 for the common-gate pair and still use the U421 for the common-source input pair. From a study of the characteristic curves for the U421, we conclude that a minimum

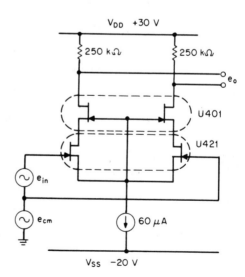

FIGURE 3-38 Cascode differential amplifier.

V_{DS} of 1 V should be allowed to ensure that it operates in drain-current saturation. This means that the minimum V_{GS} for the U401 at the operating I_D of 30 μA should be 1 V, and therefore a selection may be called for. Since the maximum $V_{GS(off)}$ for the U401 is specified as 2.5 V, the range of V_{DS} for the U421 will be 1 V to slightly less than 2.5 V. V_{GS} at $I_D = 30$ μA is specified as -1.8 V maximum for the U421; therefore, the maximum V_{DG} of the U421 for this circuit will be about 4.3 V. This is well below the I_G breakpoint. The common-mode voltage range is now limited by the breakdown rating of the U401, which is specified as 50 V. Although 50 V exceeds the I_G breakpoint of the U401, its I_G is still typically less than 10 nA; since this current is supplied by the common-source current supply, it does not contribute to input leakage. The input bias current is virtually independent of the common-mode voltage because the V_{DG} of the *input* FETs does not change with common-mode voltage.

3-11   DISTORTION IN FET AMPLIFIERS

Harmonic distortion in any amplifier stage is principally the result of a nonlinear input-output transfer characteristic plus the effect of nonlinear loading on the signal source. As the transfer curves of most amplifying devices are nonlinear, a certain amount of distortion will be generated by any amplifier, whether it employs a vacuum tube, transistor, or FET. Linearity is affected by bias, operating voltage, load impedance, signal level, and device characteristics. Input-impedance variations may also have an effect under certain operating conditions. Distortion may be minimized in any application by operating the device under the most advantageous bias and load conditions.

3-11-1   Mathematical Analysis

As certain types of distortion are more detrimental than others in any given application, an understanding of the types and causes of distortion is desirable. The type and amount of distortion observed is a function of the extent to which the transfer curve departs from a straight line. Neglecting the effects of supply voltage (except for the vacuum-tube triode) and assuming neither load-current cutoff nor saturation, transfer curves of the several amplifier devices are closely approximated as follows:

1. Triode (vacuum tube)

$$I_P = K_1 \left(V_G + \frac{V_p}{\mu} \right)^{3/2} \qquad g_m = \frac{3}{2} K_1 \left(V_G + \frac{V_p}{\mu} \right)^{1/2} \qquad (3\text{-}78)$$

2. Pentode (vacuum tube)

$$I_P = K_2 \left(V_G + \frac{V_{G2}}{\mu_{G2}} \right)^{3/2} \qquad g_m = \frac{3}{2} K_2 \left(V_G + \frac{V_{G2}}{\mu_{G2}} \right)^{1/2} \qquad (3\text{-}79)$$

3. FET

$$I_D = K_3 \left(1 - \frac{V_{GS}}{V_p} \right)^2 \qquad g_m = \frac{2K_3}{V_p} \left(1 - \frac{V_{GS}}{V_p} \right) \qquad (3\text{-}80)$$

4. Bipolar transistor

$$I_C = K_4 \left(e^{\lambda V_{BE}} - 1 \right) \qquad g_m = K_4 \lambda e^{\lambda V_{BE}} \qquad (3\text{-}81)$$

MOS-power FETs are available with a very linear region in their transfer characteristic. Although not designed for low-power small-signal amplifiers, they can be used for making low-distortion amplifiers if operated with a bias current in excess of 400 mA. This type 2N6659, for example, has a transfer curve of approximately

5. DMOS $\qquad I_D \simeq K_5 \left(V_{GS} - V_{TH'} \right) \qquad g_m = K_5 \qquad (3\text{-}82)$

K_5 for the 2N6659 is approximately 0.25 mho at values of I_D in the 400-mA to 1-A range. Below 400 mA the 2N6659 tends to follow Eq. (3-80).

To determine the degree of distortion, the transfer curves may be expressed as a power series where each term individually represents the fundamental or a distortion term. For example, the small-signal transfer function evaluated at a given set of dc operating conditions may be written as

$$i_p = g_m e_g + \frac{1}{2!} \frac{\delta g_m}{\delta V_G} e_g^2 + \frac{1}{3!} \frac{\delta^2 g_m}{\delta V_G^2} e_g^3 + \cdots + \frac{1}{n!} \frac{\delta^{n-1} g_m}{\delta V_G^{n-1}} e_g^n \qquad (3\text{-}83)$$

Note that each term includes a derivative of the transconductance term g_m. The first term represents the desired fundamental. The second represents a constant plus second harmonic, and the third represents the first and third harmonics. The first harmonic component of the third term adds to the fundamental, causing a loss in proportionality between input and output. The third-harmonic component causes cross-modulation and intermodulation distortion when two or more signals are present at the input.

With FETs, the second- and higher-order derivatives of g_m are zero; therefore, only a second harmonic is present and cross-modulation products are extremely low. In the case of vacuum tubes, fourth- and higher-order terms are negligible. However, the series does not converge so rapidly for the transistor; hence fourth- and higher-order harmonic distortion may be significant, and cross-modulation is more serious than with the vacuum tube.

As the second term of the series is proportional to curvature in the transfer characteristic, second-harmonic distortion may be minimized by operating on the most linear portion of the transfer curve.

The third term is proportional to rate of change of curvature, therefore third-harmonic and cross-modulation may be minimized in the same manner.

Although the FET promises to generate only second-harmonic distortion, the operating point must be carefully controlled for minimum distortion. Calculations made from an idealized or measured transfer curve may not necessarily tell the entire story. For example, the dc transfer curve of Fig. 3-1, normally plotted at a constant V_{DS}, does not take into account the effect of V_{DS} varying with signal or of g_{os} varying with I_D. Figure 3-5 shows both of these effects.

3-11-2 Sources of Distortion

Figure 3-39 shows a FET amplifier stage. Variations in e_{sig}, V_{GS}, R_L, $V_{GS(off)}$, g_{os}, and g_{fs} will have an effect on distortion. For low distortion, it is desirable to operate with small input signals near zero bias. Both R_L and V_{DD} should be high, and V_{DS} should be sufficiently above the knee of the output-characteristic curves but no greater than necessary. Also avoid the region near drain-to-gate breakdown to prevent the flow of gate current on signal peaks. Furthermore, certain FET geometries are preferred over others.

3-11-3 Transfer Curve and Output Conductance

Follow the instantaneous operating point up and left along the load line in Fig. 3-40. Note that g_{fs} increases as V_{GS} approaches zero. This causes second-harmonic distortion due to curvature of the transfer curve. What is not so apparent is that nonlinearities exist due to decreasing g_{os}, as I_D decreases. This effect is visible in the stylized sets of constant

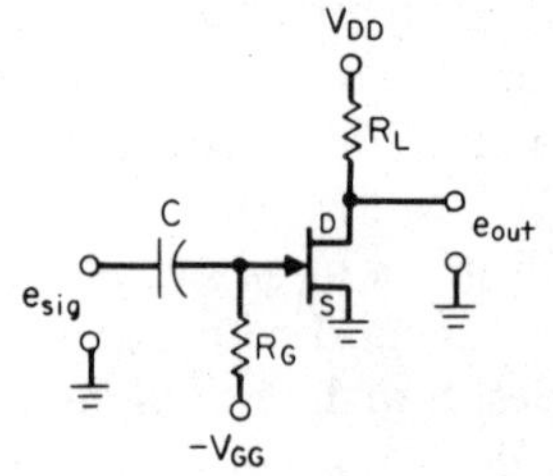

FIGURE 3-39 *RC* amplifier.

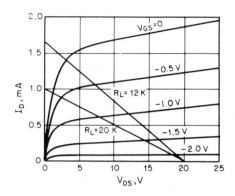

FIGURE 3-40   FET output characteristics.

g_{fs} output characteristics of Fig. 3-41. Set *a* exhibits constant g_{fs} and constant g_{os}, and set *b* exhibits a slight nonlinear change in g_{os} with I_D. Set *b* indicates an increase in gain as the operating point moves down and right along the load line. Reference to the gain equation indicates that the effect of decreasing g_{os} is less for large G_L, as expected from graphical analysis of Fig. 3-40. g_{fs} decreases with increasing V_{GS}. The decrease in g_{os} tends to offset the distortion produced by the nonlinear transfer curve. Thus at a certain operating point the two distortion sources nearly cancel to produce a point of minimum distortion.

$$\frac{e_o}{e_i} = \frac{g_{fs}}{1/R_L + g_{os}} \tag{3-84}$$

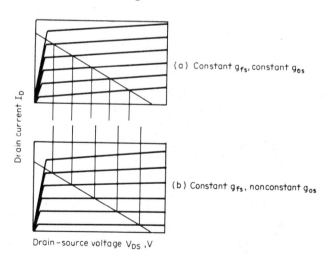

FIGURE   3-41   Idealized   FET   output   characteristics: *(a)* constant g_{fs}, constant g_{os}; *(b)* constant g_{fs}, nonconstant g_{os}.

Expanding the power series transfer function of Eq. (3-83) for the FET gives the expression

$$i_d = \frac{2I_{DSS}}{V_p^2}(V_p - V_{GS})E_a \sin \omega t - \frac{I_{DSS}}{V_p^2}E_a^2\left(\frac{\cos 2\omega t}{2}\right)\cdots \quad (3\text{-}85)$$

where E_a = peak signal amplitude.

Taking the ratio of the amplitude of the second harmonic to the fundamental, the following expression for percent second-harmonic distortion is derived:

$$\% \text{ 2d harmonic distortion} = \frac{25e_{\text{sig(pk)}}}{V_p - V_{GS}} \quad (3\text{-}86)$$

This expression is valid for small-signal distortion due only to transfer characteristic curvature. Required conditions are that the drain-voltage saturation region is avoided and that drain current flows during all portions of the cycle. Measurements of FET distortion confirm the expected results and, indeed, indicate close agreement with the distortion as calculated from Eq. (3-23).

3-11-4   Gate Bias and Signal Level

According to Eq. (3-86), small-signal distortion is directly proportional to signal level. The plot of total harmonic distortion versus input signal of Fig. 3-42 provides verification, indicating that distortion increases linearly with input signal level. Although Fig. 3-42 was generated from data taken at zero gate bias, the distortion level is unaffected by the slight forward gate biasing on negative signal peaks because the gate input impedance stays high for low values of forward bias.

Input circuit distortion does not become measurable until the rms input signal level exceeds 100 mV. This is borne out by Fig. 3-43, which plots input distortion versus input signal for gate-circuit time constants of 10 to 1000 ms. The reason for increasing distortion with input signal

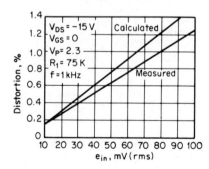

FIGURE 3-42 Distortion vs. input signal level.

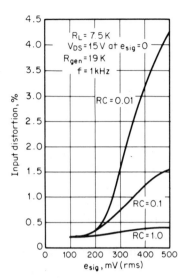

FIGURE 3-43 Distortion vs. input voltage and time constant.

is that the gate draws current from the signal source on input peaks; distortion is reduced by employing the equivalent of grid-leak bias. In this manner, the direct current drawn through the gate-return resistor develops a gate bias approximately equal to the peak forward signal voltage. So long as the input capacitor remains charged, the gate ceases to draw current on signal peaks. Thus distortion becomes a function of frequency, e.g., reducing the frequency to 100 Hz will have the same effect as reducing $T = RC$ by a factor of 10. Figure 3-43 indicates that for low distortion when the gate conducts on signal peaks, RC must be about 1000 times the period of the lowest frequency to be handled. If the generator impedance is quite low, say a few hundred ohms, the input distortion will be reduced considerably. However, as this is not a normal operating condition, the gate should be operated at a bias level sufficient to ensure that signal peaks will not forward-bias the gate by more than 100 to 200 mV.

Further reference to Eq. (3-86) indicates that distortion increases as V_{GS} approaches V_p, or that for $V_{GS} = 0$, distortion is inversely proportional to V_p. Distortion versus V_{GS} is plotted in Fig. 3-44 for two values of input signal for a device with $V_{GS(off)} = 2.3$ V. Figure 3-45 plots distortion against V_P for $V_{GS} = 0$. Rather close agreement with calculated values in evident in both plots.

3-11-5 Drain Voltage

The effects of V_{DS} and g_{os} on distortion are illustrated in Fig. 3-46. The V_{DS} effect is in reality a g_{os} effect, as may be seen from reference to

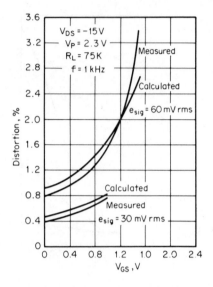

FIGURE 3-44   Distortion vs. V_{GS}.

the g_{os}-versus-V_{DS} curves of Fig. 3-47. Distortion due to g_{os} becomes large at low V_{DS} due to the very high value of g_{os}, while the distortion at high V_{DS} is principally due to the variation in g_{fs}. As V_{DS} is decreased, the g_{os}-induced distortion increases and, as previously pointed out, acts to counteract the g_{fs}-induced distortion. At a specific and fairly low V_{DS} the two out-of-phase distortions nearly cancel to produce a minimum in the distortion curve. Operation at the point of minimum distortion .is not particularly recommended, since its location is uncertain and very near a region of excessive distortion.

Operation below the point of minimum distortion at low V_{DS} results in a significant increase in nonlinear distortion due to drain saturation,

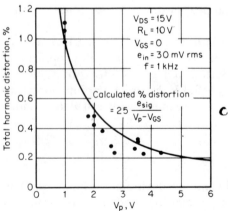

FIGURE 3-45   Distortion vs. V_P.

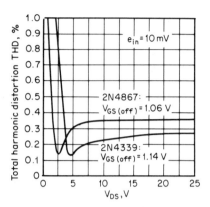

FIGURE 3-46   Distortion vs. g_{os}.

reflected in a rapidly increasing g_{os}. This effect is also visible in Fig. 3-40 as the operating point moves up and to the left along the load line approaching the knee of the output curves.

3-11-6   FET Geometry

Physical geometry of the FET has an effect on distortion, as shown in Fig. 3-46, where two devices of approximately equal $V_{GS(off)}$ generate significantly different distortion levels. These distortion levels are due to different gate lengths, which affect the value of g_{os} and the rapidity with which device output characteristics shift from triode to pentode type as V_{DS} increases. Figures 3-48 and 3-49 show the output characteristics of the two devices. The differing values of g_{os} are apparent.

The 2N4867 is a very long-channel device and exhibits the lowest value of g_{os}. The I_D transition from unsaturated to saturated is quite rapid, as shown in Fig. 3-48. Distortion is high because the low g_{os} only slightly reduces that induced by g_{fs}. On the other hand, distortion at

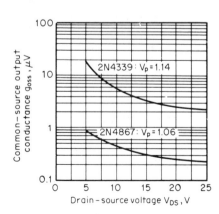

FIGURE 3-47 g_{os} vs. V_{DS}.

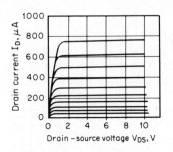

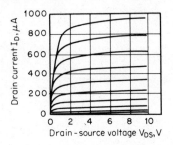

FIGURE 3-48   2N4867 output characteristics.

FIGURE 3-49   2N4339 output characteristics.

low V_{DS} is low. Where drain-supply voltages are limited, the long-gate device is preferred over other FET types because the operating point may be closer to the knee of the output curve.

The 2N4339 is a short-channel device exhibiting rather high g_{os} and a poorly defined transition region from high g_{os} to low g_{os}, as shown in Fig. 3-50. The high g_{os} results in a fairly low distortion so long as adequate drain voltage is available. The short-gate device is probably the best choice where minimum distortion is a requirement if supply voltage is not severely limited.

3-11-7   Load Resistance

The load resistance also affects distortion level. The effect is again a function of g_{os}. Figure 3-50 shows how distortion decreases with increasing R_L. The improvement is greater for short- than for long-channel devices, as g_{os} is greater for the former. The reason for an improvement with increasing R_L is that the g_{os}-induced distortion has a greater opportunity to counteract g_{fs} distortion as the load line becomes more nearly horizontal. Consider, for example, that g_{os} effects are zero when $R_L = 0$ and the load line is vertical. Conversely, g_{fs} effects are near zero when $R_L = \infty$ and the load line is horizontal. At some point with high R_L, a distortion minimum will occur beyond which g_{os} distortion is greater than g_{fs} distortion.

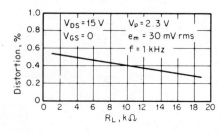

FIGURE 3-50   Distortion vs. load.

3-11-8   Rules for Low Distortion

From the preceding discussion, it follows that FET amplifiers may be operated at very low levels of distortion. However, close attention must be given to the dc operating point, bias level, load resistance, and FET characteristics. The rules to follow, listed in descending order of importance, are:

1. Maintain V_{DS} high enough that peak output signal swing will not reduce V_{DS} below 2 to 4 times $-V_{GS(off)}$.

2. Maintain V_{GS} at such a point that peak input signal swing will not forward-bias the gate junction by more than 200 mV.

3. Do not operate near drain-gate breakdown voltage unless utilizing a low signal source impedance.

4. Maintain a minimum V_{GS} consistent with other circuit requirements. More properly, maximize $V_{GS} - V_{GS}$.

5. Minimize input signal level.

6. Use a high value of load impedance.

7. Maintain V_{DS} at the lowest value consistent with rules 1 and 3.

Note that rules 4 and 6 indicate operation at the highest practical gain commensurate with power-supply and frequency-response limitations.

A particularly practical viewpoint is to consider distortion as a percentage of output voltage rather than of signal voltage as represented by Eq. (3-86). The combined effects of FET parameters $V_{GS(off)}$, I_{DSS}, g_m, and stage gain may then be determined.

Consider the stage gain expressed in Eq. (3-84) simplified to

$$\frac{e_o}{e_i} = g_m R_L \tag{3-87}$$

Combine this with Eq. (3-86) to obtain distortion in terms of output voltage.

$$\% \text{ distortion} = \frac{25e_o(pk)}{g_m R_L(V_p - V_{GS})} \tag{3-88}$$

If R_L is related to supply voltage and FET characteristics as follows:

$$R_L = \frac{V_{DD} - 2V_p}{2I_D} \tag{3-89}$$

then for maximum output signal swing, the FET characteristic equations

$$I_D = \frac{I_{DSS}}{V_p^2} (V_p - V_{GS})^2 \tag{3-90}$$

$$g_m = \frac{2 I_{DSS}}{V_p^2} (V_p - V_{GS}) \tag{3-91}$$

may be used with Eqs. (3-88) and (3-89) to find output distortion.

$$\% \text{ distortion} = \frac{25 e_{o(pk)}}{V_{DD} - 2 V_p} \tag{3-92}$$

Equation (3-92) shows that FET amplifier distortion for a given output voltage is independent of I_{DSS} and g_{fs}. Rules 4 and 5 above become of little importance for the case where e_{sig} is small and R_L may be increased without regard to bandwidth or other design considerations. High V_{DD} and low $V_{GS(off)}$ allow high R_L, high gain, and thus low e_{sig} to produce a given e_o. Thus distortion is minimized due not only to the low signal required, but also to the advantages of high R_L already discussed but not expressed in Eq. (3-92).

For lowest distortion the FET selection will depend upon the available input signal, supply voltage, and required bandwidth. For large input signal and large bandwidth, a high $V_{GS(off)}$ is desired unless source degeneration is applied. For small input signal or low V_{DD}, a low $V_{GS(off)}$ unit is desired. The choice will depend upon the specific conditions of application. In any case, a FET exhibiting high g_{oss} is desired. This characteristic may be evaluated by reference to the output characteristic curves for any particular FET. Output characteristic curves with large slope mean high g_{oss}.

Note that no attempt has been made to play off one characteristic against another. For example, a low $V_{GS(off)}$ unit from any one given geometry will exhibit a slightly higher g_{fs} at a given drain current than will the high $V_{GS(off)}$ unit of the same family. This means that the device may be operated at low I_D for higher gain g_{fs} than is attainable with the high V_p unit. Few of these tradeoffs can improve the distortion level.

BIBLIOGRAPHY

Cobbold, Richard S.C.: *Theory and Applications of Field-Effect Transistors,* Wiley, New York, 1971.

Doyle, John M.: *Digital, Switching, and Timing Circuits,* Wadsworth Publishing, Belmont, CA, 1976 [published by Duxbury Press, North Scituate, MA].

Eimbinder, Jerome: *FET Applications Handbook,* TAB Books, Blue Ridge Summit, PA, 1970.

Richman, Paul: *Characteristics and Operation of MOS Field-Effect Devices,* McGraw-Hill, New York, 1967.

Todd, Carl David: *Junction Field-Effect Transistors,* Wiley, New York, 1968.

4

HIGH-FREQUENCY CIRCUITS

4-1 HIGH-FREQUENCY TECHNIQUES

Field-effect transistors suitable for high-frequency applications have become aggressive competitors to bipolar transistors for both small-signal and power applications.

The first popular small-signal JFET suitable for VHF performance was the n-channel 2N3823 introduced by Texas Instruments in 1965. This device was characterized through 100 MHz. When Union Carbide (now Solitron) introduced their 2N4416 a year later, characterized through 400 MHz, it became the favorite for the remainder of the decade. It too had its faults, and although it maintained a lead position among high-frequency JFETs, it was not the panacea for trouble-free performance. It was close, but it missed the mark, and so there was room for the introduction of the MOSFET.

RCA introduced the first dual-gate MOSFET, the 3N140, in 1968, but not until their 3N187 came along two years later did the MOSFET become widely used in the TV and FM industry.

Both the n-channel JFET and the dual-gate MOSFET have become standard components in TV, FM, auto radios, guided missiles, radars, and wherever high-performance, high-frequency communications is found. Features and applications of some of the more prominent devices will be discussed further in this chapter.

In small-signal high-frequency applications the FET has several inherent advantages over the bipolar transistor, such as:

○ Square-law transfer characteristics which give at least a tenfold increase in input voltage handling capability with less spurious signal generation.

○ Good low-noise performance over the whole frequency range. This performance is especially noticeable in oscillators where low phase noise is of paramount importance.

○ Reduced sensitivity of input admittance to shifts in the operating point caused by AGC action, thus effecting less detuning of tuned circuits. The prime device for such applications is the dual-gate MOSFET, where 50 dB AGC can be achieved with insignificant detuning of high-Q tank circuits.

○ A high input resistance, in the common-source configuration, reduces the loading effects on input circuits, helping, among other things, to maintain high-selectivity tuned circuits.

○ Linear group delay over appreciable bandwidths and under varying AGC. Parasitic FET reactances are voltage-dependent and not generally affected by input signals.

For power FETs there are other advantages which are usually overlooked in small-signal FETs. The absence of minority-carrier storage time and of secondary breakdown is inherent in both JFETs and MOSFETs. For most bias conditions the drain current has a negative temperature coefficient; thus thermal runaway is not a likely problem. This latter characteristic simplifies the paralleling of power FETs to increase drain-current capabilities; "current hogging" does not occur. Their high-speed switching capabilities permit increased operating efficiencies in high-frequency switch-mode operation (classes D, E, and F). Their high-frequency characteristics also offer freedom from load mismatches (infinite VSWR[1] capabilities).

FETs made with gallium arsenide have made sizable encroachments

---

[1]VSWR = voltage standing wave ratio.

into the microwave bipolar transistor market. Recently developed power FETs such as the vertical-channel MOS will replace power bipolar transistors in many applications.

4-2 THE PROBLEM OF HIGH-FREQUENCY AMPLIFICATION

It was not until the manufacturers had refined their processes to achieve FETs with high transconductances and low interelectrode and parasitic capacitances that there emerged a suitable high-frequency device.

As the operating frequency is increased, the effects of parasitic capacitances become the fundamental restriction upon the FET—in particular, the effect on its input impedance. In the common-source stage, the Miller effect becomes the dominant capacitance. Input impedance is largely a reactance of C_{in}:

$$X_{in} = \frac{1}{2\pi f C_{in}} \qquad (4\text{-}1)$$

where $C_{in} = C_{iss} - A_V C_{rss}$

For the common-source amplifier the quantity A_V is negative (due to the phase reversal); therefore

$$\begin{aligned} C_{in} &= C_{iss} + |A_V C_{rss}| \\ &= C_{gs} + C_{gd} + |A_V C_{dg}| \end{aligned} \qquad (4\text{-}2)$$

The reverse capacitance C_{rss} of a FET, in addition to raising the input capacitance through the Miller effect, feeds back signals from the output to the input, contributing to possible instability—especially if the load itself is reactive.

Much of the design effort in high-frequency amplifiers is directed toward reduction of the effects of the feedback capacitance. Although the common-gate amplifier has lower power gain than the common-source type, it is widely used at high frequencies because of its much lower feedback capacitance. The magnitude of the voltage gain for these two types of amplifiers may be comparable; however, in the common-source stage, C_{gd} is the feedback capacitor, while in the common-gate stage it is C_{ds}. Also A_V for the common gate is positive; thus, for the common gate,

$$C_{in} = C_{gs} + C_{ds} - |A_V| C_{ds} \qquad (4\text{-}3)$$

For many device types C_{ds} is an order of magnitude lower than C_{gd}. The resulting reduction in the feedback capacitance makes the common-gate amplifier attractive despite its lower power gain. Other advantages will be examined in this chapter.

A common figure of merit used to compare high-frequency performance of FETs is

$$\text{Figure of merit (FM)} = \frac{g_m}{2\pi C_{\text{in}}} \qquad (4\text{-}4)$$

and for the popular n-channel JFET, the U310, this figure equates to 600 MHz. To illustrate the improvement in n-channel JFETs over the period of 1965 to 1976, the following should be of interest:

Type	FM relative to 2N4416
2N3823	0.6
2N4416	1.0
2N5397	1.3
U310	2.1

There are expressions of high-frequency performance other than the classic figure of merit [Eq. (4-4)] that should be considered. Although perhaps not immediately apparent, a "quality factor" expression

$$\text{QF} = \frac{g_{fs}}{g_{is}} \qquad (4\text{-}5)$$

becomes especially important in the selection of JFETs suitable for ultra-broadband transmission-line (distribution) amplifiers.

DESIGN PRIORITIES FOR HIGH-FREQUENCY AMPLIFIERS

Amplification of	Analog signals	Digital signals
Small signal	High intercept	Flat group delay
	Low crossmods	High intercept
	Low noise figure	Low crossmods
	Sufficient gain	Low noise figure
	AGC	Sufficient gain
		AGC
Large signal	Linear	
	Low spurious	
	Low sideband noise	Same
	Overload protected	
	Mismatched loading	

In small-signal amplifiers an equally important ratio is that of forward transconductance to output conductance, more properly identified as mu.

$$mu = \frac{g_{fs}}{g_{os}} \tag{4-6}$$

Each of these quality factors may serve as a guide in the selection of the proper transistor for a given application. Seldom does one achieve the perfect FET. What a "perfect" FET might be will be reviewed later in this chapter. Design priorities for high-frequency amplifiers can be outlined according to applications (see table on page 136).

FETs more closely meet these design objectives than do any other known active semiconductor. This chapter will continue to examine these priorities as they are applied to amplifiers, mixers, and oscillators.

4-3 THE "PERFECT" HIGH-FREQUENCY FET

There is no "perfect" high-frequency FET, nor is there likely to be one. Factors that inhibit optimum high-frequency gain and noise figure are: (1) feedback capacitance C_{gd}; (2) shunting input conductance g_{is}; and (3) output conductance g_{os}. High forward transconductance g_{fs} is of paramount importance, and its magnitude is directly affected by the magnitude of both the input and output conductances.

In the common-source amplifier the input conductance g_{is} poses a mixed problem. Although a very low input conductance raises the quality factor (g_{fs}/g_{is}), which is desirable, a more debilitating effect occurs. A very low conductance (a high input resistance) infers high input Q, and effective coupling from the input circuitry may well result in excessive losses and increased overall noise figure in wideband circuits.

Output conductance g_{os}, on the other hand, directly affects the forward gain of the FET. Maximum available voltage gain is fundamentally related to the ratio

$$\frac{g_{fs}}{g_{os}}$$

An appreciation of how g_{fs} must be improved to overcome the debilitating effects of both the shunting effects of increasing g_{is} and the reduced mu caused by the increasing g_{os} is provided in Fig. 4-1.

The chart in Fig. 4-1 is interpreted as follows: P_{GU} is unilateralized power gain expressed in decibels. But note carefully that each value is underlined. For example, 10 dB (solid underline) and 15 dB (dashed underline).

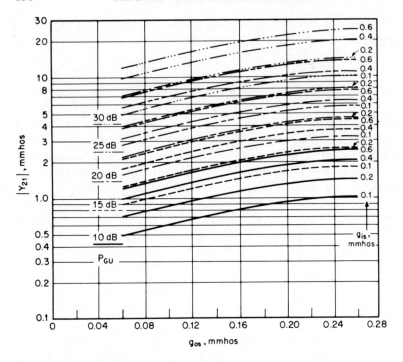

FIGURE 4-1 Projected UHF JFET transconductance as a function of g_{is}, g_{os}, and gain. Unilateralized power gain common source ($Y_{12} = 0$).

The value of g_{is} (right vertical axis) ranges from 0.1 to 0.6 mmho and is repeated five times. Again, note that each time a value of g_{is} is repeated, it is to identify a different coded graph that corresponds to the underlined value of P_{GU}.

For example, assume we desire a unilateralized power gain P_{GU} of 20 dB. We have a value of g_{is} of 0.4 mmho and a value of g_{os} of 0.12 mmho. From this chart we determine that we need a value of $|Y_{21}|$ of 4.3 mmhos.

Feedback affects circuit stability, complicates circuit design, and reduces performance. Manufacturers reduce C_{gd} by careful attention to the metalizations and diffusions in the design of the FET chip as well as in chip packaging.

4-4 STABILITY

A very important consideration in the design of a small-signal, high-frequency amplifier is its stability, its freedom from a tendency to oscillate. Amplifier "stability factors" have been derived by Linvill[2] and by

[2]Carson, Ralph S., "High Frequency Amplifiers," Wiley Interscience, New York, 1975.

Stern.[3] Linvill assumes that the amplifier stage has infinite source and load impedance, which are the worst-case stability conditions. Stern takes into account not only the transistor, but also the source and load impedances.

The Linvill stability factor C is expressed as

$$C = \frac{|Y_f Y_r|}{2\text{Re } Y_1 \text{Re } Y_0 - \text{Re}(Y_f Y_r)} \qquad (4\text{-}7)$$

If C is less than 1, the transistor is considered to be unconditionally stable. If C is greater than 1, oscillation is possible if the transistor is presented with some specific source and load impedance; it is considered to be only conditionally stable. Unconditional stability assures that any positive real impedance presented to the transistor will not cause it to oscillate. This assumes no external feedback.

The Stern stability factor K is

$$K = \frac{2(g_{11} + g_s)(g_{22} + g_1)}{|Y_{12} Y_{21}| + \text{Re}(Y_{12} Y_{21})} \qquad (4\text{-}8)$$

In this expression it is assumed that the input and output matching networks are simultaneously tuned to the same frequency. The Stern solution defines $K > 1$ as being unconditionally stable. If g_s and g_1 (source and load conductance) are set equal to zero, the result is the reciprocal of the Linvill stability factor!

There are many ways to ensure stability in an amplifier. From the equations for the Linvill stability factor the following equations can be derived:

$$g_{\text{input loading}} = \frac{|Y_{21} Y_{12}|/C + \text{Re}(Y_{21} Y_{12})}{2 g_{22}} - g_{11} \qquad (4\text{-}9)$$

$$g_{\text{output loading}} = \frac{|Y_{21} Y_{12}|/C + \text{Re}(Y_{21} Y_{12})}{2 g_{11}} - g_{22} \qquad (4\text{-}10)$$

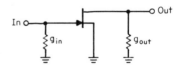

FIGURE 4-2   Two-port with conductance loading.

$g_{\text{input loading}}$ and/or $g_{\text{output loading}}$ are those values of conductance which must be added to the transistor externally, as in Fig. 4-2, to obtain the desired stability factor C.

In many cases it is not desirable to use these techniques to stabilize

[3]Ibid.

a transistor. Some of the power which would ordinarily be delivered to the load is now dissipated in the stabilizing resistors.

Stern developed a technique in which the input and output could be mismatched by equal amounts to assure stability of the amplifier. This technique causes a mismatched input and output as the price for circuit stability. In this case one establishes the Stern stability factor and then calculates by an iterative process the values for b_{in} and b_{out}.

$$G_s = \sqrt{\frac{K[|Y_{21} Y_{12}| + \text{Re}(Y_{21} Y_{12})]}{2}} \sqrt{\frac{g_{11}}{g_{22}}} - g_{11} \qquad (4\text{-}11)$$

$$G_1 = \sqrt{\frac{K[|Y_{21} Y_{12}| + \text{Re}(Y_{21} Y_{12})]}{2}} \sqrt{\frac{g_{22}}{g_{11}}} - g_{22} \qquad (4\text{-}12)$$

where one can solve for both b_{in} and b_{out} (the *susceptance* values of Y_{in} and Y_{out}) by

$$Y_{in} = Y_{11} - \frac{Y_{21} Y_{12}}{Y_{22} + Y_1} \qquad (4\text{-}13)$$

and

$$Y_{out} = Y_{22} - \frac{Y_{21} Y_{12}}{Y_{11} + Y_s} \qquad (4\text{-}14)$$

A third method of ensuring stability is by neutralization or unilateralization. Neutralization is a technique used to set b_{12} or the reactive component of the feedback y_{12} equal to zero. In most instances this would be equivalent to adding, between the input and output of the transistor, an inductor whose reactance is equal in magnitude to the reactance of the transistor feedback capacitance (C_{rss}). This is equivalent to placing a parallel tuned trap between the input and output matching networks.

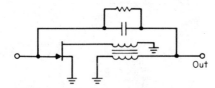

FIGURE 4-3 Two-port with unilaterization feedback.

Unilateralization is that technique in which g_{12} and b_{12} are set to zero. This might be implemented as shown in Fig. 4-3, by using a 1:1 phase-inverting transformer at the input or output with a resistor and capacitor equal to g_{12} and b_{12} tied back across the transistor.

The advantages of neutralization and unilateralization are that the input and output are effectively isolated from each other and maximum stable gain results. The disadvantage is that in most instances (among small-signal amplifiers) the operating Q is increased, forcing a decrease in bandwidth.

Most JFETs and single-gate MOSFETs, when operating in their common-source configuration, are only conditionally stable ($C > 1$). Because they are conditionally stable, the precautions for stability must be observed. Alternatives to the common-source configuration are the common-gate and the Cascode configurations. When a JFET is operated in the common-source configuration, there is a relatively large intrinsic capacitance C_{gd} between input and output. However, when the JFET is used common-gate, C_{gd} is shunted to ground, and capacitance between input and output is the relatively small drain-to-source capacitance, generally in femtofarads.

An effective Cascode arrangement is the popular dual-gate MOSFET, where the isolation between input and output can reach 50 dB, thus affording considerable stability.

4-5 POWER GAIN

Power gain may be defined as the ratio of power delivered to the load to power absorbed at the input of the amplifier. Therefore, the power gain of a JFET common-source amplifier should be very high, as the input resistance is high, thus absorbing little power. Conversely, one can expect the common-gate amplifier to have reduced gain compared with its common-source counterpart. However, in the common-source configuration, because of stability requirements, the gain of the common-source JFET amplifier must be reduced.

There are many things to consider when designing an amplifier for a specific power gain. More often than not, system gain requirements may be less than the gain capability of the transistor. For example, a required stage gain of 10 dB is needed from a JFET that could easily offer 15 dB.

Given the expression for transducer gain[4]

$$G_t = \frac{4g_{11}g_{22}\,|Y_{21}|^2}{|(Y_{11} + Y_s)(Y_{22} + Y_1) - Y_{12}Y_{21}|^2} \tag{4-15}$$

the power gain of an amplifier can be determined. G_t (transducer power gain) is the gain of an amplifier when the input and output matching networks, as well as the transistor, are considered. In general, gain can be reduced by lowering the transconductance of the FET, or the load impedance or the source impedance.

[4]Ibid.

FET gain is generally attributed to its forward transconductance. As mentioned in earlier chapters, a FET has two fundamental operating areas: the triode area, where $r_{DS(on)}$ predominates, and the pentode area of current saturation. Controlling transconductance by the manipulation of current is tedious and not recommended. When the need to lower the gain arises, it is often beneficial to manipulate the input impedance to bring it closer to the source impedance and thus provide a better match.

If the transistor is unconditionally stable, maximum available gain (MAG) is

$$\text{MAG} = \frac{|Y_{21}|^2}{A + (A^2 - |Y_{12}Y_{21}|^2)^{1/2}} \tag{4-16}$$

where $A = 2g_{11}g_{22} - \text{Re}(Y_{21}Y_{12})$

To obtain MAG both the input and output must be a conjugate match. For these conditions,

$$g_s = \frac{1}{2g_{22}}(A^2 - |Y_{21}Y_{12}|^2)^{1/2} \tag{4-17}$$

$$b_s = -b_{11} + \frac{\text{Im}(Y_{21}Y_{12})}{2g_{22}} \tag{4-18}$$

$$g_1 = \frac{1}{2g_{11}}(A^2 - |Y_{21}Y_{12}|^2)^{1/2} \tag{4-19}$$

and
$$b_1 = -b_{22} + \frac{\text{Im}(Y_{21}Y_{12})}{2g_{11}} \tag{4-20}$$

where $A = 2g_{11}g_{22} - \text{Re}(Y_{21}Y_{12})$

There are many situations where G_{max} is greater than it is required to be. In such cases, Stern offers a technique by which the input and output are equally mismatched to obtain a desired stability factor and a gain less than MAG. Linvill has described a technique where only the output is mismatched and the input is conjugately matched. Both Stern and Linvill lend themselves to graphical solutions.

4-6 TWO-PORT TRANSISTOR PARAMETERS

4-6-1 Z, Y, H, and S Parameters

Transistors are characterized at high frequency using one of four possible parameter sets. Z (impedance) parameters correspond to open-circuit constraints on the unexcited terminal pair. Y (admittance) parameters correspond to a short circuit at the unexcited terminal pair.

H (hybrid) parameters are a mixture of Z and Y parameters. The H parameter set is popular for bipolar transistor characterization because of the widely divergent impedances of bipolar transistors. In particular, for the common-base and common-emitter configurations, it is impractical to characterize the transistor with either Z or Y parameters. Consequently, H parameters have different dimensions.

S (scattering) parameters suggest that an incident signal impinging upon a mismatched load scatters into different components. If so, then S parameters provide a measure of this separation of scattered elements. Recently high-frequency design has focused heavily on S parameters rather than the formerly popular Y parameters. Both S and Y will be briefly discussed in this chapter. Z parameters and H parameters, which are more popular among bipolar transistor designers, will be omitted.

4-6-2 Y Parameters

Short-circuit admittances are still widely used in high-frequency calculations. The equations describing the common-source equivalent circuit of Fig. 4-4 are

$$I_{in} = Y_{is}E_{in} + Y_{rs}E_{out} \tag{4-21}$$

$$I_{out} = Y_{fs}E_{in} + Y_{os}E_{out} \tag{4-22}$$

where E_{in} and I_{in} are the input signal voltage and current and E_{out} and I_{out} are the output signal voltage and current.

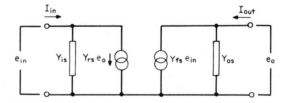

FIGURE 4-4   High-frequency admittance parameter JFET model.

Short-circuiting the output ($E_{out} = 0$) yields

$$Y_{is} = \frac{I_{in}}{E_{in}} \tag{4-23}$$

and

$$Y_{fs} = \frac{I_{out}}{E_{in}} \tag{4-24}$$

Short-circuiting the input ($E_{\text{in}} = 0$) yields

$$Y_{rs} = \frac{I_{\text{in}}}{E_{\text{out}}} \qquad (4\text{-}25)$$

and

$$Y_{os} = \frac{I_{\text{out}}}{E_{\text{out}}} \qquad (4\text{-}26)$$

thus

$$\begin{vmatrix} Y_{is} & Y_{rs} \\ Y_{fs} & Y_{os} \end{vmatrix} = Y_{is} Y_{os} \qquad (4\text{-}27)$$

since

$$\frac{I_{\text{in}}}{E_{\text{in}}} \times \frac{I_{\text{out}}}{E_{\text{out}}} = \frac{I_{\text{out}}}{E_{\text{in}}} \times \frac{I_{\text{in}}}{E_{\text{out}}} \qquad (4\text{-}28)$$

then

$$Y_{is} Y_{os} - Y_{rs} Y_{fs} = 0$$

Figure 4-5 shows a neutralizing path added to the FET. Adding this to the FET Y parameters, we arrive at

$$\begin{vmatrix} Y_{is} & Y_{rs} \\ Y_{fs} & Y_{os} \end{vmatrix} + \begin{vmatrix} Y_n & -Y_n \\ -Y_n & Y_n \end{vmatrix} = \begin{vmatrix} Y_{is} + Y_n & Y_{rs} - Y_n \\ Y_{fs} - Y_n & Y_{os} + Y_n \end{vmatrix} = 0$$

so

$$(Y_{is} + Y_n)(Y_{os} + Y_n) = (Y_{rs} - Y_n)(Y_{fs} - Y_n) \qquad (4\text{-}29)$$

and

$$I_{\text{in}} = (Y_{is} + Y_n) E_{\text{in}} + (Y_{rs} - Y_n) E_{\text{out}} \qquad (4\text{-}30)$$

$$I_{\text{out}} = (Y_{fs} - Y_n) E_{\text{in}} + (Y_{os} + Y_n) E_{\text{out}} \qquad (4\text{-}31)$$

Since in neutralizing we equate $Y_{rs} = Y_n$, then

$$I_{\text{in}} = (Y_{is} + Y_{rs}) E_{\text{in}} + 0 \qquad (4\text{-}32)$$

$$I_{\text{out}} = (Y_{fs} - Y_{rs}) E_{\text{in}} + (Y_{os} + Y_{rs}) E_{\text{out}} \qquad (4\text{-}33)$$

Since I_{in} is unaffected by E_{out}, neutralizing is effective in cancelling all feedback effects. In practice, Y_{rs} is largely capacitive, making the neutral-

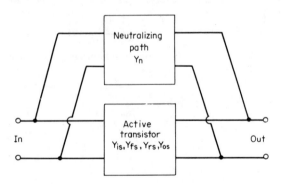

FIGURE 4-5   Basic two-port with feedback neutralization.

izing path largely inductive. Both paths—through the FET and through the neutralizing path—must have the same reactance at the frequency for which the amplifier is to be designed. That is,

$$2\pi fL = \frac{1}{2\pi fC_{gd}} \tag{4-34}$$

or

$$L = \frac{1}{4\pi^2 f^2 C_{gd}} \tag{4-35}$$

Each Y parameter is a complex quantity and may be derived from a "physical" equivalent circuit somewhat more comprehensive than that shown for low frequencies.

Y parameters are functions of frequency and are so described in most high-frequency data sheets. Typical of such graphs are those of Fig. 4-6, which relate to the Siliconix U310 JFET. Notice that each Y parameter is specified in terms of its real and imaginary components. Such data are most useful for computer-aided design (CAD).

Many FET data sheets show Y parameters extending to 1 GHz. However, in the conventional TO46, 52, or 72 package, performance above 600 MHz is limited because of the package capacitances and internal lead-bond inductances.

4-6-3 S Parameters

Scattering or S parameters are receiving increased attention in the characterization of devices and the design of high-frequency circuits. Modern computer-aided design (CAD) optimization programs, such as COMPACT,[5] rely upon the designer "inputting" S-parameter data. Some simplified S-parameter designs have omitted the important reverse transfer parameter S_{12} by equating it to zero ($S_{12} = 0$); although this greatly simplifies the "number crunching" of two-port network design, the results often have little utility.

When either the input scattering parameter S_{11} or the output scattering parameter S_{22} is displayed graphically, frequently superimposed on a Smith chart, one must be careful to remember that these scattering parameters are *reflection coefficients*, not impedance. Although both reflection coefficients and impedances can be plotted on a Smith chart, *they are by no means the same.*

Since the basic S parameter is derived from either voltage or current,

---

[5]COMPACT = *C*omputerized *O*ptimization of *M*icrowave *P*assive and *A*ctive *C*ircui*T*s. COMPACT Software, Inc., Patterson, NJ 07504.

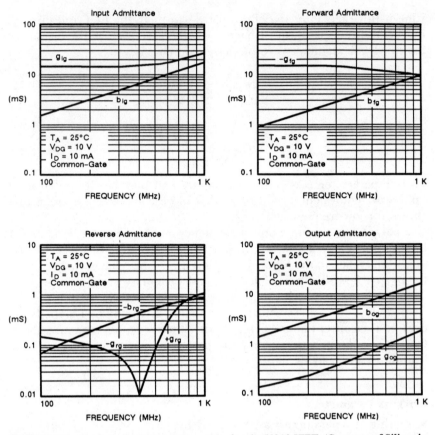

FIGURE 4-6 Typical admittance parameters for the U310 JFET. (Courtesy of Siliconix Incorporated.)

it is correct to identify power parameters as the square of the S parameter. Consequently, for symmetrical networks the following identifications are noteworthy.

$$\% \text{ power reflected} = |S_{11}|^2 \times 100 \qquad (4\text{-}36)$$

$$\% \text{ power transmitted} = |S_{21}|^2 \times 100 \qquad (4\text{-}37)$$

$$\% \text{ power dissipated} = (1 - |S_{11}|^2 - |S_{21}|^2) \times 100 \qquad (4\text{-}38)$$

A logical step shows that the forward power gain of a transistor is

$$G_{(\text{dB})} = 10 \log |S_{21}|^2 \qquad (4\text{-}39)$$

The forward operating power gain, taking into account reflections from the input port, is

$$G_{P(dB)} = 10 \log \left(\frac{|S_{21}|^2}{1 - |S_{11}|^2} \right) \qquad (4\text{-}40)$$

Because of the increasing popularity and growing importance of S parameters, it is advisable for the reader to consult the massive literature on this subject.

4-7 AMPLIFIERS

The requirement for low cross modulation and intermodulation products makes the FET a practical necessity in solid-state RF circuitry. Recent improvements in the design of gallium arsenide FETs have extended their useful operating frequency into the gigahertz region, and the more conventional silicon JFETs to above 500 MHz. The Y parameters given in Fig. 4-6 are indicative of modern JFET performance at VHF/UHF.

The JFET can offer noise performance superior to many other forms of active amplifying elements, particularly at low frequencies. Of special interest is the effect of $1/f$ noise as related to RF amplifiers, which results in low phase jitter and low AM noise sidebands.

JFET amplifier design may use any of several possible configurations. The common source, the common gate, and the common drain (source follower) are the most universal. Each has its advantages and disadvantages. The Cascode design, being a combination of both the common source and common gate, offers distinct advantages. Each will be covered in this section.

4-7-1 The Common-Source Amplifier

In the common-source configuration (Fig. 4-7), the JFET exhibits several advantages, such as high input impedance, which, in turn, contributes to high gain. A low noise figure is achieved with a noise-impedance match nearly equal to the power-gain impedance match. The major disadvantage of this common-source amplifier is that it is often potentially unstable, thus requiring special design precautions (see the section on stability).

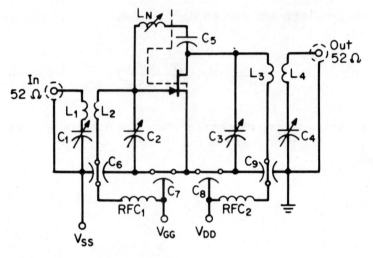

FIGURE 4-7 450-MHz neutralized CS amplifier (C_{1-4}—0.8 to 12 pF Johanson type 2950; C_2—40 pF DM5 silver mica; C_{6-9}-1000 pF Allen-Bradley type FA5C; L_1—1.1 inch long, 16 gauge solid copper; L_3—1.3 inch long, 16 gauge solid copper; L_4—1.4 inch long, 22 gauge enamel, spaced 0.3 inch from L_3; RFC$_{1,2}$—0.15 μHy Delevan type 1537-00; L_N—3 turns, 22 gauge enamel, 0.25 inch ceramic form, aluminum slug).

An additional disadvantage of using the common-source configuration is its susceptibility to cross-modulation products which are voltage-dependent. Since the input impedance of the common-source amplifier is high, a signal of known input power will produce a high voltage on the gate, giving rise to potentially serious cross-modulation products.

The high input impedance of the common-source configuration allows a voltage step-up transformer to be used between the low-impedance line and the input to the JFET gate. Gain due to this voltage transformation may be expressed

$$A_V = \sqrt{\frac{R_{\text{in}}}{R_{\text{gen}}}} \tag{4-41}$$

If care is not taken, the low noise figure that may be expected in this circuit configuration will be lost due to the insertion loss inherent in the input network. A ratio of unloaded Q to loaded Q should be greater than 10:1 to minimize this loss.

$$\text{Insertion loss (dB)} = -20 \log \left(1 - \frac{Q_l}{Q_u} \right) \qquad (4\text{-}42)$$

In Fig. 4-7 both the input and output circuits are capacitively tuned. At the input the interelectrode capacitance C_{iss} contributes significantly to the tuned network.

There are practical difficulties involved with neutralizing. Each circuit must be carefully adjusted for maximum gain at the desired frequency, and during the neutralizing procedure the signal fed back into the output port must be of a comparable magnitude, as will be anticipated in the output circuit during normal operation. Each amplifier must be individually neutralized for the specific JFET, as generally no two are alike. Shielding between input and output must be used for optimum performance. Quite often the success of the common-source neutralized amplifier hangs on the placement of this shielding!

4-7-2 The Common-Gate Amplifier

The common-gate amplifier is the most popular high-frequency JFET amplifier because:

○ The low input resistance $(1/g_m)$ together with the low input capacitance offers convenient matching to transmission lines and other low-impedance sources.

○ The low input resistance offers excellent immunity to cross modulation.

○ It is unconditionally stable, generally through UHF, due to low feedback.

Yet there are disadvantages; for example,

○ Optimum noise impedance match does not offer optimum power gain.

○ It cannot accept AGC without severe output-circuit detuning (not much different than any other JFET configuration).

○ It has lower power gain than common source.

In many situations the JFET used in the common-gate (CG) configuration is unconditionally stable even at rather high frequencies (Fig. 4-8). However, care must be exercised in the design and layout of the CG amplifier; in particular, shielding between source and drain loads is critical. A

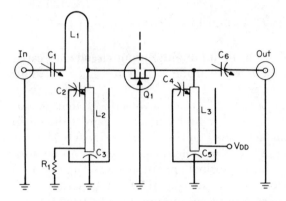

PARTS LIST FOR U310, 450-MHZ TEST FIXTURE

1.	Socket*	Augat	8060–1G9
2.	$L_2 L_3$**	NIBCO	Copper coupling 600, 1 in
3.	$C_1 C_2 C_4$	Johanson Manufacturing	0.8–10 pF, 5200
4.	$C_3 C_5$	Allen Bradley	1000-pF feedthrough
5.	C_6	Erie	0.25–1.5 pF, 530–000
6.	RF connector (two)	King	UG-1094
7.	L_1	—	14 awg bare copper, 2.8 in long with ⅜-in diameter bend, uniform lead length.
8.	Chassis	BUD	CU234

* Socket was divided to accept a copper foil shield soldered directly to both the gate lead and ground.
** Center conductor is ¼-in diameter copper tubing.

FIGURE 4-8   Test fixture U310—450 MHz.

JFET showing complete stability can become an oscillator if, for example, a low-loss socket is used without interelectrode shielding being maintained *within the socket itself!* A few femtofarads of coupling between socket pins can bring instability to an otherwise unconditionally stable amplifier.

The common-gate and common-source configurations have a common fault: their high drain impedance, which severely limits their performance over wide bandwidths. There are several solutions to achieve wide bandwidths, but none that does not limit the overall stage gain of the JFET amplifier. One easy solution is to load the drain circuit either by mismatch or by resistive loading; another means used in multi-stage amplifiers is stagger tuning; still another, but of little practical application in modern design, is the distributed amplifier design.

The common-gate amplifier, shown in Fig. 4-8, not having a frequency-dependent neutralization path, not only is free from the relevant adjustment problems, but is effective as a wideband amplifier. However, because of the lower input impedance, it does not offer the power gain of the common-source amplifier.

4-7-3 Common-Drain Amplifier (Source Follower)

The common-drain amplifier (sometimes called the source follower or voltage follower) characteristically has high input and low output impedances. It is frequently used to couple high-impedance video lines to low-impedance loads and is typically used at frequencies below 100 MHz.

The real part of the output impedance is $(g_{fs})^{-1}$, which is generally independent of frequency. If the load resistance is greater than $(g_{fs})^{-1}$, then C_{in} is also independent of frequency.

The high-frequency corner f_3 (where the response is down 3 dB) is an inverse function of the input time constant $R_g C_{in}$:

$$f_3 = \frac{1}{R_g C_{in}} \qquad (4\text{-}43)$$

where

$$C_{in} = C_{gd} + C_{gs}(1 - A_V) + C_{stray} \qquad (4\text{-}44)$$

and R_g is the input signal source resistance. Examples of how the signal source resistance affects bandwidth is shown in Fig. 4-9.

The voltage gain of the source follower is slightly less than unity and is given by

$$A_V = \frac{g_{fs} R_s}{1 + g_{fs} R_s} \qquad (4\text{-}45)$$

From this we observe that the voltage gain is nearly independent of R_s when the $g_{fs} R_s$ product is large.

The output resistance of the FET, g_{fs}^{-1}, is in *parallel* with the source resistor R_S. Therefore, R_O of the amplifier is

$$R_O = \frac{R_S}{1 + g_{fs} R_S} \qquad (4\text{-}46)$$

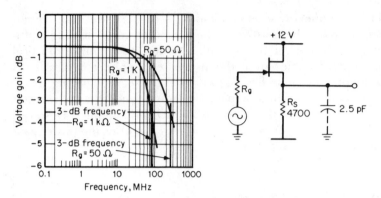

FIGURE 4-9 Effect of source impedance on bandwidth of a video amplifier.

4-7-4 The Cascode Amplifier

The FET Cascode circuit shown in Fig. 4-10 brings to solid-state amplifiers the high-frequency characteristics of the pentode vacuum tube, that is, reduced Miller effect resulting in improved stability and bandwidth. Although JFETs may be used in Cascode circuits, they have pretty much given way to the more popular dual-gate MOSFET, which is a monolithic Cascode element.

The Cascode amplifier, whether it is constructed using a pair of JFETs or a single dual-gate MOSFET, provides a tenfold reduction in feedback capacitance compared with a simple single-stage common-source JFET amplifier.

Practical advantages of this reduced Miller effect can be demonstrated in the gain-bandwidth calculations for an *RC*-coupled video amplifier. The gain-bandwidth product (GBW) is defined by

$$\text{GBW} = \frac{g_{fs}}{2\pi(C_{\text{in}} + C_{\text{out}})} \tag{4-47}$$

For the single common-source JFET, C_{in} may be calculated:

$$C_{\text{in}} = C_{gs} + C_{dg}(1 - A_V) \tag{4-48}$$

and

$$C_{\text{out}} = C_{dg} \tag{4-49}$$

Here the Miller-effect capacitance, $(1 - A_V)C_{dg}$, places a shunting input capacitance proportional to the stage gain. This limits high-frequency performance. In the Cascode amplifier stage the effective loading of the second stage upon the drain connection of the first stage limits the first stage gain to near unity (−1) and thereby forces the input capacitance C_{in} to be

$$C_{in} = C_{gs} + 2C_{dg} \qquad (4\text{-}50)$$

C_{iss} is $C_{gs} + C_{dg}$; consequently, $C_{in} = C_{iss} + C_{dg}$. The improvement of the gain-bandwidth product between a JFET and a dual-gate MOSFET can be most effectively shown by comparing the Siliconix U311 with the 3N201, both exhibiting near-equal forward transconductance g_{fs}.

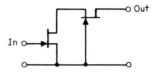

FIGURE 4-10 Basic Cascode configuration.

Operating the U311 in the common-source configuration, the typical parameters are:

$$g_{fs} = 14 \text{ mmhos}$$
$$C_{gs} = 3.2 \text{ pF}$$
$$C_{gd} = 1.2 \text{ pF}$$

The 3N201 parameters are:

$$g_{fs} = 14 \text{ mmhos}$$
$$C_{gs} = \ 4 \text{ pF}$$
$$C_{gd} = 0.2 \text{ pF (including strays)}$$

Beginning with the U311 with a load resistance of 2000 Ω,

$$A_V = g_{fs}R_1 = 14 \times 10^{-3} \times 2 \times 10^3 = 28$$
$$C_{in} = C_{gs} + C_{gd}(1 + A_V) = 3.2 \times 10^{-12} + (1.2 \times 10^{-12})(1 + 28) = 38 \text{ pF}$$
$$C_{out} = C_{gd} = 1.2 \times 10^{-12} \text{ F}$$

From Eq. (4-47),

$$\text{GBW} = \frac{14 \times 10^{-3}}{6.28(38 \times 10^{-12} + 1.2 \times 10^{-12})} = 56.8 \text{ MHz}$$

For the dual-gate MOSFET 3N201,

$$A_V = g_{fs}R_1 = 28$$
$$C_{in} = C_{gs} + 2C_{gd} = 4 \times 10^{-12} + 2(0.2 \times 10^{-12}) = 4.4 \text{ pF}$$
$$C_{out} = C_{gd} = 0.2 \times 10^{-12}$$

$$\text{GBW} = \frac{14 \times 10^{-3}}{6.28(4.4 \times 10^{-12} + 0.2 \times 10^{-12})} = 485 \text{ MHz}$$

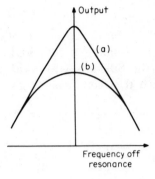

FIGURE 4-11 Amplifier response: *(a)* without feedback and *(b)* with inverse feedback.

The contrast is unmistakably in favor of the dual-gate MOSFET when GBW is of paramount importance. An additional benefit when using the dual-gate MOSFET is the simplicity of applying AGC by varying the bias on the second gate. Where this would severely modify the parasitic capacitances of a simple JFET, the dual-gate MOSFET, because of its greatly reduced feedback capacitance (and thus lower Miller effect), maintains a near-constant capacitance, thus reducing any detuning effects that would naturally occur with AGC control on the simple JFET.

Gain-bandwidth product can also be improved by the use of inverse feedback that is a maximum at band center and drops off toward the edges of the band. As shown in Fig. 4-11, a flattening of the response curve of the amplifier with a consequent increase in the bandwidth results. If the increase in bandwidth is greater than the decrease in gain, the GBW is increased.

4-7-5   The Broadband Amplifier

Undoubtedly the fundamental shortcoming of FETs as amplifiers has been their high output resistance (low output conductance), which has limited their application in broadband applications. The video source-follower amplifier is the notable exception since, being a *source* follower, its output impedance is effectively $1/g_{fs}$ in parallel with the source resistor R_s. However, the other amplifier configurations, common source, common gate, and Cascode, are generally plagued with a high drain impedance.

Yet there are solutions. Perhaps the simplest that offers increased high-frequency response, particularly in video amplifier applications, is the shunt peaking technique, where a small inductor is placed in series with the load resistor. The value of inductance for maximum flat-gain response may be calculated

$$L = \frac{R_1^2(C_{in} + C_{out})}{2} \tag{4-51}$$

Generally broadband techniques, other than the shunt-peaking method, involve Q manipulation of the output tuned circuit. This becomes especially critical when multistage amplifiers are built simply because of the bandwidth shrinking effect common to *synchronously tuned* stages:

$$BW_{overall} = BW_{one\,stage}[\sqrt{(2^{1/n} - 1)}] \qquad (4\text{-}52)$$

Perhaps the easiest and most popular method of improving bandwidth of the common-gate amplifier is to load the output to reduce Q. Once the overall gain and bandwidth have been established, the number of stages can be closely approximated and calculations made to determine the proper loading.

4-7-6 Amplifier Design Example

The following example should be of benefit to the reader in understanding the procedure.

Let an amplifier be designed for a 225-MHz center frequency with a 1-dB bandwidth of 50 MHz and 24 dB power gain using the Siliconix U310 JFET. Some familiarity with the U310 suggests that three stages are necessary. Solving for a single-stage bandwidth using the bandwidth shrinking formula, Eq. (4-52), the single-stage bandwidth for 50 MHz, 1 dB passband is 98 MHz. To achieve this bandwidth requires a loaded tuned circuit in the drain of each FET. The required Q is determined by

$$\frac{BW}{f_0} = \frac{1}{Q}\sqrt{\left(\frac{E_0}{E}\right)^2 - 1} \qquad (4\text{-}53)$$

Solving for Q gives

$$Q = 1.15$$

Referring to the admittance parameters of the U310 for a center frequency of 225 MHz, a close estimate of the drain load admittance equates to a drain output impedance of 3700 Ω shunted by approximately 2.5 pF (includes some stray capacitance). To achieve the required *single-stage* bandwidth,

$$Q = \omega C R_t \qquad (4\text{-}54)$$

where R_t is that value of load impedance required to achieve the single-stage bandwidth.

$$R_t = \frac{1.15}{1.415 \times 10^9 \times 2.5 \times 10^{-12}} = 330\ \Omega \qquad (4\text{-}55)$$

Since the JFET itself exhibits a drain load resistance of 3700 Ω, the net resistance that must be placed across the tank circuit is about 365

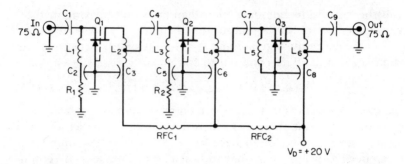

FIGURE 4-12   Schematic of a three-stage synchronously tuned bandpass amplifier ($C_1, C_4, C_7, C_9 = 68$ pF; $C_2, C_5 = 500$ pF; $C_3, C_6, C_8 = 1000$ pF; $Q_1, Q_2, Q_3 = $ Siliconix U310; $L_1, L_3, L_5 = 120$ nH; $L_2, L_4, L_6 = 222$ nH; $RFC_1, RFC_2 = 2.2$ μH; $R_1, R_2 = 51$ Ω).

Ω. This, of course, can be made a part of the tank circuit design. The circuit shown in Fig. 4-12 exhibits the response shown in Fig. 4-13.

Although this procedure (and those to follow) suggests that bandwidths can be achieved with FETs, one must not forget the limiting gain-bandwidth product (GBW), which remains inviolable. Consequently, there is a maximum bandwidth possible at a given overall gain. For synchronously tuned stages, shown in the example, maximum bandwidth occurs when the stage gain reduces to 4.34 dB (10 log e).

Other broadbanding schemes are available, and most are affected in some manner by bandwidth shrinking when multiple stages are used. It is, however, possible to reduce or even to eliminate bandwidth shrinkage by using over- and undercoupled multituned circuits, which have been well covered in the literature.

Another less popular method of improving bandwidth is the distributed amplifier. In the distributed amplifier, rather than terminating the drain in a tank circuit, the parasitic capacitances of the FET contribute

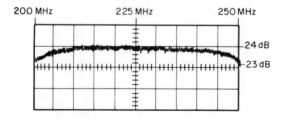

FIGURE 4-13   Passband response of Fig. 4-12.

to the necessary shunting transmission-line capacitance of the common-drain output network. Wide bandwidths are possible; however, the complexity of the design negates its popularity. One of the most debilitating effects for GBW is the gate conductance at high frequencies [or quality factor—see Eq. (4-5)].

4-8 MIXERS

There has been a great deal of interest in JFET mixers because they offer the opportunity of both high dynamic range and conversion gain. At one time it was thought that lower noise figures could also be attained, but experience has shown that with commercially available JFETs a noise figure comparable to that of the Schottky-barrier diode is, at best, difficult.

Perhaps the most favorable aspect of mixer design using JFETs is the square-law transfer characteristic. The JFET mixer has lower intermodulation and harmonic products than does a comparable bipolar or diode mixer. Active JFET mixers operating at high levels outperform passive mixers in terms of dynamic range and large signal-handling capabilities.[6]

Additionally, the active mixers offer improved conversion efficiencies over their passive counterparts. This permits relaxation of the IF amplifier gain requirements and in some cases the elimination of the RF preamplifier. The basic large-signal model, shown in Fig. 4-14, is considered to have an ideal square-law transfer characteristic, with a parasitic resistor R_s, in series with the source. The shunting capacitances are shown to be equal, as most FETs are closely symmetrical with respect to their gate.

To incorporate the effect of noise into the FET model, it is necessary to identify the sources of noise. Van der Ziel has shown that the primary sources of noise in a FET are bulk resistance in series with the drain and source, and channel-width modulation by thermal noise in the conducting channel; thus

$$\overline{i_n^2} = 4KTB\gamma g_{fs} \tag{4-56}$$

where $\gamma = 0.6$

A mixer circuit accepts a high-frequency modulated input signal, mixes it with a high-frequency unmodulated signal from a local oscillator, and produces a series of frequencies which are functions of the original

---

[6]It is also true that Cascoding diode sets has produced outstanding dynamic range and signal handling, but at the expense of vastly increased local oscillator power.

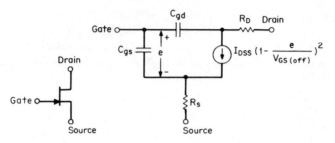

FIGURE 4-14 FET large-signal model ($C_{gs} = C_{gd} = C_o/[1 + (d/V)]^\alpha$; $C_{gs} = C_o/(d + V_{gs})^\alpha$; $C_{gd} = C_o/(d + V_{gd})^\alpha$; α = **law of capacitance variation for junctions**; d = **work voltage for junction, usually 0.75 V**; C_o = **junction capacitance at 0 V**; I_{DDS} = **drain current for $e = 0$**; V_R = **gate-to-source voltage for drain** OFF **current equal to 1 μA). (Reprinted with permission from D. M. Hodson, "UHF Fet Mixer of High Dynamic Range," U.S. Army ECOM Research and Development Technical Report #ECOM-0503-P005-G821, December 1969.)**

two. One of these, usually the difference frequency, is selected by tuned circuits and is referred to as the "intermediate frequency" (IF). The IF, which is modulated with the same waveform as the incoming signal, is then amplified and the modulation extracted for whatever use the system requires. Mixers are generally used for frequency conversion, generally to a lower frequency for more effective handling, but quite often to higher frequencies where filtering can be more effective in removing unwanted sidebands (or images).

Initial evaluation of the active JFET mixer compared to bipolar mixers will imply a disadvantage because of local oscillator drive requirements; bipolar devices in low-level mixers require very little drive power. However, in high-level mixing this disadvantage is overcome because drive requirements are generally about equal.

Since JFETs have transfer characteristics approximating a square-law response, their third-order intermodulation distortion products are generally much smaller than those of bipolar transistors. Harmonic distortion and cross-modulation effects are third-order-dependent, and thus are greatly reduced when FETs are used in active balanced mixers.

A secondary advantage derives from available conversion gain, so that the FET mixer becomes simultaneously equivalent to both a converter and a preamplifier.

Design criteria, in order of priority, include the following:

1. Intermodulation and cross modulation

2. Conversion gain

3. Noise figure

4. Selecting the proper FET

5. Local oscillator injection

6. Designing the input transformer

7. Designing the IF network

4-8-1 Intermodulation and Cross Modulation

A basic aim in mixer design is to avoid the effects of intermodulation product distortion and cross modulation. Part of the problem may be resolved by using a balanced mixer circuit.

The active transfer function of the FET is represented by a voltage-controlled current source. For both cross modulation and intermodulation, the amount of distortion is proportional to the amplitude of the gate-source voltage. Since input power is proportional to input voltage and inversely proportional to input impedance, the best FET IM and cross-modulation performance is obtained in the common-gate configuration, where the impedance is lowest.

When JFETs are used as active mixer elements, it is important that the devices be operated in their square-law region. Operation in the FET square-law region will occur with the device in the depletion mode. Considerable distortion will result if the JFET is operated in the enhancement mode (positive V_{GS} for an n-channel JFET). The problems encountered are similar to those which arise when positive drive is placed on the grid of a vacuum tube.

Square-law-region operation emphasizes the importance of establishing proper drive levels for both quiescent bias and the local oscillator. The maximum conversion transconductance g_c is achieved at about 80 percent of the FET gate cutoff voltage $V_{GS(off)}$, and amounts to about 25 percent of the forward transconductance g_{fs} of the FET when used as an amplifier. Since conversion gain (or loss) must be considered, it is common to equate voltage gain A_V as

$$A_V = g_c R_L \tag{4-57}$$

where g_c is the conversion transconductance and R_L is the FET drain load.

An attempt to achieve maximum conversion gain by indiscriminately increasing the drain-load resistance will adversely affect any design priority concerning distortion—particularly intermodulation-product distortion.

Distortion takes different forms in mixers. Most obvious is that distortion which will occur if the FET is driven into the enhancement mode, as noted earlier. Finally, there is the so-called varactor effect.

The most frequent cause of poor mixer performance stems from signal overloading in the drain circuit. Excessive drain-load impedance degrades the intermodulation characteristics and produces unwanted cross-modulation signals.

A characteristic of the FET balanced mixer is that the correct drain-load impedance is inversely proportional to the value of the conversion transconductance. Figure 4-15 shows the improvement in IM characteristics obtained in the prototype mixer with the drain-load impedance reduced to 1700 Ω from 5000 Ω. Specifically, the dynamic load line must be plotted so that the signal peaks of the instantaneous peak-to-peak output voltage are not permitted to enter into the nonsaturated region of the FET. Suitable and unsuitable drain-load lines are shown in Fig. 4-16. Load impedance selection is quantified in Eqs. (4-72) through (4-74).

Distortion from the varactor effect is of secondary importance; it arises from an excessive peak voltage signal swing, where the changing drain-to-source voltage can cause a change in parasitic capacitance C_{rss} and give rise to harmonics.

A FET tends to be voltage-dependent when the drain voltage falls appreciably below 6 V. If the source voltage (from the power supply) is also low and the drain-load impedance is high, then distortion will develop. However, if proper steps are taken to prevent drain-load distortion, the varactor effect will also be inhibited.

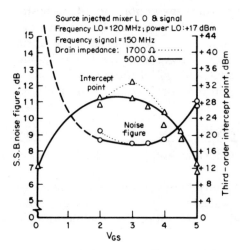

FIGURE 4-15 Comparison of mixer IM characteristics.

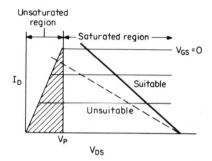

FIGURE 4-16 **Plotting drain load lines.**

4-8-2 Conversion Gain

In a FET, forward transconductance is defined as

$$g_{fs} = \frac{dI_D}{dV_{GS}} \tag{4-58}$$

and conversion transconductance is defined as

$$g_c = \frac{dI_{D(\omega i)}}{dV_{GS(\omega r)}} \tag{4-59}$$

where ωi = intermediate frequency
       ωr = signal frequency

The effects of time-varying local oscillator voltage v_2 and the much smaller signal voltage v_1 must be considered:

$$v_{GS} = v_1 \cos \omega_1 t + v_2 \cos \omega_2 t \tag{4-60}$$

For square-law operation,

$$V_2 + V_{GS} \leqslant V_{GS(off)} \tag{4-61}$$

Drain current is approximately defined by

$$I_D = I_{DSS} \left(1 - \frac{V_{GS}}{V_{GS(off)}} \right)^2 \tag{4-62}$$

or

$$I_D \approx \frac{g_{fso} V_{GS(off)}}{2} \left(1 - \frac{v_{gs}}{V_{GS(off)}} \right)^2 \tag{4-63}$$

or

$$I_D \approx \frac{g_{fso}}{2 V_{GS(off)}} (V_{GS(off)} - v_{gs})^2 \tag{4-64}$$

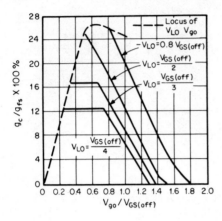

FIGURE 4-17   Normalized g_c/g_{fs} vs. $V_{GS}/V_{GS(off)}$. (Reprinted with permission from S.-P. Kwok, "Field Effect Transistor RF Mixer Design Techniques," WESCON/67 *Convention Record*, Session 8.)

then

$$I_D \approx \frac{g_{fso}}{2V_{GS(off)}} \qquad \text{(complex Taylor expansion)} \qquad (4\text{-}65)$$

which can be reduced to

$$I_{D(IF)} \approx \frac{g_{fso}}{2V_{GS(off)}} V_1 V_2 \cos(\omega_1 - \omega_2)t \qquad (4\text{-}66)$$

and the conversion transconductance is

$$g_c = \frac{g_{fso}}{2V_{GS(off)}} |V_2| \qquad (4\text{-}67)$$

Equation (4-67) suggests that g_c increases without limit as V_2 increases without limit. However, to avoid operation of the FET in the unsaturated region, the peak-to-peak swing of V_2 should not exceed $V_{GS(off)}$. Thus

$$2V_2 \text{ peak} \leq V_{GS(off)} \qquad (4\text{-}68)$$

or

$$V_2 \text{ peak} \leq \frac{V_{GS(off)}}{2} \qquad (4\text{-}69)$$

Figure 4-17 shows plots of normalized conversion transconductance, g_c/g_{fs}, versus normalized quiescent bias, $V_{GS}/V_{GS(off)}$, for different oscillator injections.

4-8-3   Noise Figure

Like the common-gate FET amplifier, the common-gate FET balanced mixer is sensitive to generator resistance R_g. A change of a decade in R_g can produce a noise-figure variation of as much as 3 dB.

4-8-4 How to Select the Proper FET

Conversion efficiency is determined by conversion transconductance g_c, which in turn is directly related to zero-bias saturation current I_{DSS} and the gate cutoff voltage $V_{GS(off)}$:

$$g_c = \frac{I_{DSS}}{V_{GS(off)}^2} |V_2| \qquad (4\text{-}70)$$

$$\approx \frac{g_{fso}}{2V_{GS(off)}} |V_2| \qquad (4\text{-}71)$$

Equation (4-71) appears to indicate that FETs with high I_{DSS} are to be preferred. However, I_{DSS} and $V_{GS(off)}$ are related, and Figs. 4-18a and b show that devices from a family selected for high I_{DSS} do *not* provide high conversion transconductance, but actually produce a lower value of g_c.

Basic considerations in selecting FETs for this application are gate cutoff voltage $V_{GS(off)}$ for good conversion transconductance, and zero-bias saturation current I_{DSS} for dynamic range. Among currently available devices, the U310 offers excellent performance in both categories.

There is, of course, the possibility that FET cost is a major consideration in evaluating the active balanced mixer approach—the familiar price/performance tradeoff. If this is the case, there are a number of other FETs which will provide suitable alternatives to the U310. Remember, however, that conversion transconductance g_c can never be more than 25 percent of forward transconductance. Thus, as tradeoff considerations begin, the first sacrifice to be made will be the degree of achievable conversion gain. Intermodulation performance will follow, with the third tradeoff being available noise figure. Table 4-1 lists a number of possible alternatives to the U310.

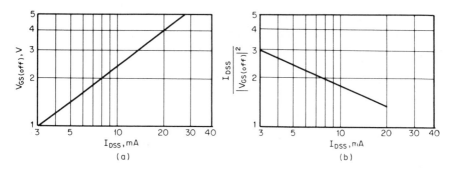

FIGURE 4-18 Relationship of I_{DSS} and $V_{GS(off)}$.

TABLE 4-1 COMPARISON OF JFET PARAMETERS CRITICAL TO MIXER PERFORMANCE

Typical characteristic	Device type		
	U310	2N4416	2N3823
g_m	14 kΩ	5 kΩ	3.5 kΩ
I_{DSS}	40 mA	10 mA	10 mA

4-8-5 Designing the IF Network

The IF network performs two important functions in the FET mixer circuit. It provides for optimum match between the FET and the IF amplifier at the IF frequency, and it effectively bypasses signal and local oscillator frequencies.

In network design, it is essential that the RF and local oscillator signals be sufficiently isolated from the intermediate-frequency signal to maintain rejection levels of at least 20 dB. If this isolation is not maintained, conversion gain and noise figure are degraded.

A first-order approximation to establish the proper load impedance may be obtained when

$$R_L = \frac{V_{DD} - 2V_{GS(off)}}{i_d} \tag{4-72}$$

where

$$i_d = I_{DSS}\left(1 - \frac{v_{gs}}{V_{GS(off)}}\right)^2 \tag{4-73}$$

and

$$v_{gs} = V_{GS} + V_1 \sin \omega_1 t \tag{4-74}$$

For the U310 FET, the optimum drain-load impedance is established at slightly less than 2000 Ω, with sufficient local oscillator drive and gate bias determined from the conversion transconductance curve in Fig. 4-17.

4-9 BALANCED MIXERS

When high-performance, high-frequency junction FETs are used in active balanced mixers, the performance is superior to that obtained with hot-carrier diodes. Several types of mixers are compared in Table 4-2. The advantages and disadvantages of semiconductor devices currently used in various mixer circuits are shown in Table 4-3.

TABLE 4-2 COMPARISON OF MIXER TYPES

Characteristic	Mixer type		
	Single-ended	Single-balanced	Double-balanced
Bandwidth	Several decades possible	Decade	Decade
Relative IM density	1.0	0.5	0.25
Interport isolation	Little	10–20 dB	>30 dB
Relative LO power	0 dB	+3 dB	+6 dB

TABLE 4-3 COMPARISON OF SEMICONDUCTOR DEVICES FOR MIXERS

Device	Advantages	Disadvantages
Bipolar transistor	Low noise figure High gain Low dc power	High IM Easy overload Subject to burnout
Diode	Low noise figure High power handling High burnout level	High LO drive Interface to IF Conversion loss
JFET	Low noise figure Conversion gain Excellent IM products Square-law characteristic Excellent overload High burnout level	Optimum conversion gain not possible at optimum square-law response level High LO power
Dual-Gate MOSFET	Low IM distortion AGC Square-law characteristic	High noise figure Poor burnout level Unstable

4-9-1 Why FETs for Balanced Mixers?

Modern communication systems have the stringent requirements of wide dynamic range, suppressed of intermodulation products, and low cross modulation. These parameters must be considered before noise figure and gain are taken into account.

4-9-2 First-Order Single Balanced Mixer Theory

Essential details of balanced mixer operation, including signal conversion and local oscillator noise rejection, are best illustrated by signal-flow vector diagrams (Fig. 4-19).

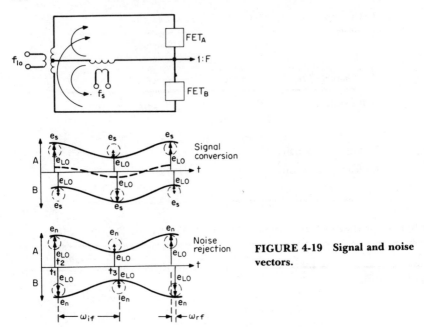

FIGURE 4-19   Signal and noise vectors.

Energy conversion into the intermediate-frequency (IF) passband is the major concern in mixer operation. In the following analysis, both the signal and noise vectors are shown progressing (rotating) at the IF rate (ω_{if}); the resulting wave occurs through vector addition.

The analysis of local oscillator noise rejection (Fig. 4-19) assumes, for simplicity of explanation, that noise is coherent. Thus at some point in time t_1 the noise component e_n is "in phase" with the local oscillator vector e_{lo}, and FET A (the rectifying element) is ON; the JFET mixer acts as a switch, with the local oscillator acting as the switch drive signal. One-half cycle later, at time t_2, the signal flow is reversed for both the local oscillator vector and the noise component; FET A is OFF and FET B is ON. Moving ahead an additional one-half of the IF cycle, FET A is again ON, but the noise component has advanced 180° (ω_{if}) through the coupling structure, and is now "out of phase." The process continually repeats itself.

The end result of this averaging (detection) is the cancellation of the noise which originated in the local oscillator, provided that the mixer balance is precise.

The analysis of the conversion of the signal to the IF passband is similar, but the signal is injected into the coupling structure at the equipotential tap. Thus at time t_2, the signal vector e_s is "out of phase"

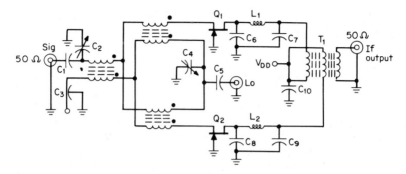

FIGURE 4-20 Prototype active balanced mixer [C_1, C_5—0.01 μF; C_2, C_4—1 to 10 pF; C_3—1000 pF; C_6, C_8—30 pF; C_7, C_9—68 pF; C_{10}—0.1 μF; L_1, L_2—1.3 μH; Q_1, Q_2—U310 (2) or U430; T_1—RELCOM BT-9].

with the local oscillator vector e_{lo}. The resulting envelope develops a cyclic progression at the IF rate, since the signal is "demodulated" by the mixing action of the FETs.

A schematic of a *prototype* balanced mixer is shown in Fig. 4-20. In the design of the prototype FET active balanced mixer, the generator resistance of the FETs is established by the hybrid coupling transformer. Two important criteria for the FETs in the circuit are high forward transconductance and a value of power-match source admittance g_{igs} which closely matches the output admittance of the coupling transformer. In the common-gate configuration, match points for optimum power gain and noise do not occur at the same value of generator resistance, as shown in Fig. 4-21. Optimum noise match can only be achieved at the sacrifice of bandwidth.

Best mixer performance is achieved with "matched pairs" of JFETs. Basic considerations in selecting FETs for this application are gate cutoff voltage $V_{GS(off)}$ for good conversion transconductance, and zero-bias saturation current I_{DSS} for dynamic range. A match to 10 percent is generally adequate. Among currently available devices, the Siliconix U310 and the dual U431 offer excellent performance in both categories; common-

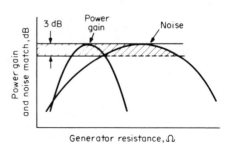

FIGURE 4-21 Power gain and noise matching.

gate forward transconductance is 20,000 μmhos max at $V_{DS} = 10$ V, $I_D = 10$ mA, and $f = 1$ kHz.

4-9-3 Local Oscillator Injection

Low IM distortion products and noise figure, plus best conversion gain, will be achieved if the voltage swing of the local oscillator across the gate-to-source junction is held to the values presented in Fig. 4-17. V_{LO} is expressed in terms of peak-to-peak voltage, while $V_{GS(off)}$ is a dc voltage.

Local oscillator injection can be made either through a brute-force drive into the JFET source through the hybrid input transformer, or through a direct-coupled circuit to the JFET gates, where less drive will be required for the desired voltage swing. Two circuits to obtain direct gate coupling are shown in Fig. 4-22.

The source-injection method is used in the design of the present mixer to maintain the inherent stability of a common-gate circuit. A minor disadvantage with the direct-drive method is that the required gate-to-source voltage swing requires considerable local-oscillator input power. For source injection through the transformer, best mixer performance is obtained with a local-oscillator drive level of +12 to +17 dBm across a 50-Ω load.

Direct coupling to the FET gates would occur at a higher impedance level, which would result in less local oscillator drive power. However, when the gates are tied together, shunt susceptance requires some form of conjugate matching. This brings about an undesirable reduction of instantaneous mixer bandwidth.

The performance of the active mixer is clearly superior to that of diode mixers, as shown in Fig. 4-23 and Table 4-4; it contributes to

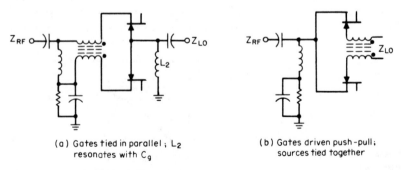

(a) Gates tied in parallel; L_2 resonates with C_g

(b) Gates driven push-pull; sources tied together

FIGURE 4-22 Alternate forms of LO injection.

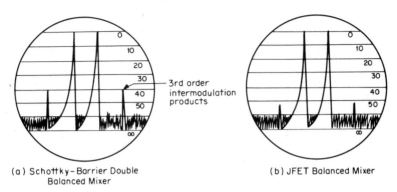

(a) Schottky-Barrier Double
Balanced Mixer

(b) JFET Balanced Mixer

FIGURE 4-23   Comparison of third-order IM products.

overall system gain in areas critical to telecommunications practice and reduces associated amplifier requirements.

4-10   DOUBLE-BALANCED MIXERS

The comparison in Table 4-2 clearly shows those performance characteristics of the double-balanced mixer which have made it the most popular of all mixer types. Among these attributes are greatly improved interport isolation and a significant degree of rejection of local oscillator carrier amplitude modulation.

Passive devices, however, such as Schottky-barrier (hot-carrier) diodes, have certain fundamental shortcomings, such as high conversion

TABLE 4-4   50 TO 250-MHz MIXER PERFORMANCE COMPARISON

Characteristic	JFET		Schottky		Bipolar	
Intermodulation intercept point	+32	dBm	+28	dBm	+12	dBm*
Dynamic range	100	dB	100	dB	80	dB*
Desensitization level (the level for an unwanted signal when the desired signal first experiences compression)	+8.5	dBm	+3	dBm	+1	dBm*
Conversion gain	+2.5	dB†	−6	dB	+18	dB
Single-sideband noise figure @ 50 MHz	7.2	dB	6.5	dB	6.0	dB

* Estimated.

† Conservative minimum.

loss and high local oscillator drive requirements. The active balanced mixer which employs FETs is a welcome innovation. Conversion gain and improved intermodulation distortion characteristics alone place the FET double-balanced mixer far ahead of its passive counterparts. The high saturation levels possible with modest local oscillator power make such a mixer useful for mixing both small and large signals.

Double-balanced mixers using MOSFETs have been considered; however, the MOSFETs were used solely as switching devices, requiring no external dc power. As a result, MOSFET mixers have exhibited high conversion loss and require considerable local oscillator drive power.

4-10-1   First-Order Double-Balanced Mixer Theory

In either single- or double-balanced mixer design, the prime requirement is that when the mixer is excited by the local oscillator carrier, the circuit must be capable of rejecting the amplitude-modulated wave which exists about the LO. Also, the mixer must reject any AM signal entering from the local oscillator port. (This signal rejection is usually known as AM local oscillator noise cancellation.)

A second requirement for balanced mixers is the establishment of interport isolation between the signal, local oscillator, and IF ports. A third desirable characteristic is the reduction of intermodulation distortion products. Careful attention to the design of double-balanced mixers will satisfy the foregoing criteria.

Figure 4-24 shows the schematic of a JFET double-balanced mixer in what this author identifies as a "transconductance mixer." In Sec. 4-11 the merits of a commutation (switching) mixer are discussed. The four high-performance junction FETs are chosen for closely matched characteristics. The significance of the quad-FET configuration will be dealt with later.

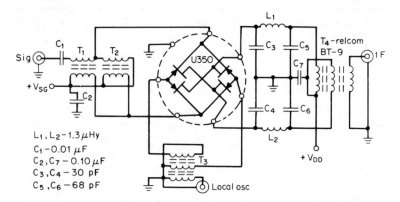

FIGURE 4-24   Double-balanced mixer.

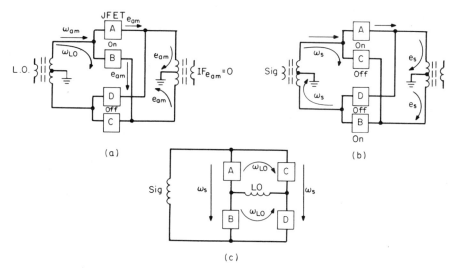

FIGURE 4-25   Double-balanced mixer analysis.

The schematic of Fig. 4-25a is simplified to show only the local oscillator circuit so that the rejection mechanism of AM signals, either on
the LO carrier or entering through the local oscillator port, is easier
to understand. Likewise, the equivalent circuit in Fig. 4-25b demonstrates
how the signal is enhanced at the IF output. Local oscillator AM cancellation and signal enhancement are dependent upon the precise balance
of the IF transformer, as well as on the match of the four FETs which
make up the quad network. In Fig. 4-25c, the schematic has been re-
arranged to show both the local oscillator and the signal input transformers; the mechanics of interport isolation may be easily visualized.
Signal excitation provides an equipotential at the junctions of the local
oscillator transformer and FET pairs AB and CD; in the same manner,
excitation of the local oscillator produces an equipotential balance at
the junctions of the signal transformer and FET pairs AC and BD.

Harmonic distortion products are reduced by the balance between
the signal and local oscillator (inputs) and the IF (output), where even-
integer harmonics of the signal and local oscillator frequencies are effectively cancelled. A sixth-order summary of such products in both single-
and double-balanced mixers is shown in Table 4-5. Note how the relative
densities agree with Table 4-2. The effects of harmonic distortion can
be reduced by a judicious selection of the IF passband response. Third-
order IMD (intermodulation distortion) products are reduced by virtue
of the characteristics of junction FETs, which approximate a square-
law response. Care must be taken in FET operation, however, to avoid

TABLE 4-5 COMPARISON OF MODULATION
PRODUCTS IN SINGLE- AND DOUBLE- BAL-
ANCED MIXERS TO SIXTH ORDER

Single-balanced	Double-balanced
f_s	
$3f_s$	
$5f_s$	
$f_0 \pm f_s$	$f_0 \pm f_s$
$f_0 \pm 3f_s$	$f_0 \pm 3f_s$
$f_0 \pm 5f_s$	$f_0 \pm 5f_s$
$2f_0 \pm f_s$	
$2f_0 \pm 3f_s$	
$3f_0 \pm f_s$	$3f_0 \pm f_s$
$3f_0 \pm 3f_s$	$3f_0 \pm 3f_s$
$4f_0 \pm f_s$	
$5f_0 \pm f_s$	$5f_0 \pm f_s$

driving the device into forward conductance by the application of too
much local oscillator power.

4-10-2 Harmonic Distortion, Intermodulation Products, and Cross Modulation

Spurious output signals in mixers fall into three categories:

1. Spurious mixer products derived from harmonic mixing
   of the signal and local oscillator frequencies

2. m-Tone, n-order intermodulation products

3. "Chirping," which arises from undesired mixing frequencies falling in the IF passband

The harmonics of a single-signal frequency, when mixed with the harmonics of the local oscillator, produce spurious outputs which are level-
dependent on the signal amplitude. These products are greatly reduced
by the double-balanced mixer, where the even harmonics are effectively
cancelled; when FETs are used, the Taylor-series power expansion falls
quickly to zero above the second order.

However, modulation products of a similar nature will arise if the
broadband down-converting mixer is not preceded by signal preselec-
tion, because of the mixer's equal response to the "image" frequency.
Here, perfectly valid signals will mix with the local oscillator, producing
interfering IF signals whose only difference, when compared to the de-

sired IF signal, is that they move counter to the desired IF signal when the local oscillator is shifted.

Two-tone, odd-order IM products differ markedly from other spurious signals. This form of harmonic distortion consists of interactions between two or more input signals and their respective harmonics. In turn, these products are mixed with the fundamental and harmonics of the local oscillator, generating spurious products which may fall within the IF passband, on or very near to the desired signal.

Cross modulation in the active JFET balanced mixer does not pose a serious problem provided the signal input is maintained at a high conductance, which will occur with source injection. Cross modulation is very dependent on and directly related to the impedance across which the signal is impressed. In the active JFET double-balanced mixer this impedance is very low, typically 35 Ω. Consequently, the effects of cross modulation may be disregarded.

In the mixing process of any active device, the value of the FET drain current may be derived from a knowledge of the transconductance of the device and the impressed signal voltage e_g. This is obtained from the Taylor-series power expansion:

$$i_d = g_m e_g + \frac{1}{2!}\frac{\partial g_m}{\partial V_G}e_g^2 + \frac{1}{3!}\frac{\partial^2 g_m}{\partial V_G^2}e_g^3 + \cdots + \frac{1}{n!}\frac{\partial^{n-1}g_m}{\partial V_G^{n-1}}e_g^n \qquad (4\text{-}75)$$

which can be broken down into the components shown in Table 4-6.

In FET theory, the second- and higher-order derivatives of g_m are absent, and the device thus offers a considerable reduction of both intermodulation products and higher-order harmonics. In the double-balanced mixer, where $F1 \pm F2$ is the desired result, it is well to manipulate mixer design and bias conditions to render $\partial g_m/\partial V_G$ as large as possible, simultaneously reducing all other terms.

TABLE 4-6   ELEMENTS OF THE TAYLOR POWER SERIES

Term	Output	Transfer characteristic
$g_m e_g$	F1, F2	Linear
$\dfrac{1}{2!}\dfrac{\partial g_m}{\partial V_G}e_g^2$	2F1, 2F2 F1 ± F2	Second-order square-law
$\dfrac{1}{3!}\dfrac{\partial^2 g_m}{\partial V_G^2}e_g^3$	3F1, 3F2 2F1 ± F2 2F2 ± F1	Third-order

4-10-3   Local Oscillator Injection

Local oscillator drive for active FET mixers, either balanced or unbalanced, differs from the drive characteristics of passive diode mixers. In the switching mode, the diode mixer requires sufficient local oscillator drive to swing the diodes from a hard ON state to a hard OFF state. For best IMD performance, the gate of the FET must never be driven positive with respect to the source—a case equivalent to the hard ON condition of the diode. Consequently, local oscillator drive for the balanced mixer is less than that required for a passive balanced mixer *with comparable performance characteristics.*

The double-balanced mixer relies on balanced drive from both the local oscillator and the signal source. Since conversion efficiency, optimum noise figure, and good cross-modulation effects can best be served with the signal entering through the common quad JFET source, the local oscillator excitation may be applied directly at the gates of the FET array.

4-10-4   AM Local Oscillator Noise Rejection

Originally, balanced mixers were used for the specific purpose of cancelling spurious AM signals existing on or about the local oscillator carrier (the function of the mixer in establishing good interport isolation was a side effect). These signals could be either spurious AM signals generated on or about the carrier (Fig. 4-26) or actual signals existing at the signal frequency. In the latter case, the signals enter the mixer through the local oscillator, having found their way in through some leakage coupling phenomenon.

Regardless of the type or source of AM signals entering through the local oscillator port, the balanced mixer should effectively reject these signals so that their products do not occur at the intermediate

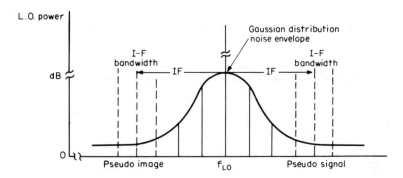

FIGURE 4-26   Generation of spurious AM signals.

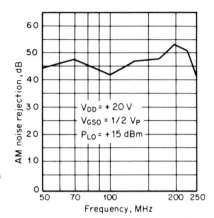

FIGURE 4-27 AM noise rejection in double-balanced mixer.

frequency. In the early days of balanced mixers, a 20-dB rejection of AM noise was considered good; today's sophisticated techniques for selection of dynamically matched semiconductors can provide AM rejection in excess of 30 dB. Figure 4-27 provides an insight into the degree of AM noise rejection available in the double-balanced mixer. (Insofar as FM noise is concerned, it should be noted that no mixer is capable of rejecting frequency-modulated signals entering through the local oscillator.)

An interesting point not generally considered in discussions of balanced mixers is that the dynamic range of the mixer is limited by the conversion of local oscillator noise into the intermediate frequency. This blanks out a weak signal and places a bottom on sensitivity.

4-10-5 Interport Isolation

Like AM noise rejection and dynamic unbalance, interport isolation is very dependent on mixer balance (symmetry). Matching aspects of the JFET quad array and the phase/amplitude balance of the signal input and local oscillator input transformers play important roles in achieving interport isolation. Capacitive and magnetic coupling between the transformers add to problems of interport isolation in balanced mixers.

Interport isolation was enhanced in the prototype mixer through careful parts layout. As a measure of the overall effects of unbalance, a quantitative measurement of interport isolation versus dynamic unbalance is made in Fig. 4-28. Dynamic unbalance may be regarded as another expression for AM noise rejection, except that the latter does not provide a ready insight into the effects of symmetry, balance, and quad match.

In Fig. 4-29, the interport isolation between the local oscillator and signal input ports is shown to be 35 dB typically.

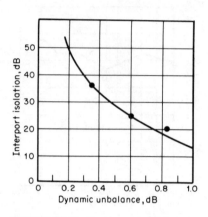

FIGURE 4-28   Interport isolation vs. dynamic unbalance.

FIGURE 4-29   Interport isolation.

Selection of the dynamic drain impedance in the IF network is a critical point in the design. Both IM product distortion and cross modulation will be affected by the instantaneous peak-to-peak voltage of the FETs if the dynamic drain impedance allows the signal peaks to enter either the pinchoff or breakdown-voltage regions of the transistors. Here another design tradeoff must be considered. If the impedance is too high, the dynamic range of the mixer will be limited; if the impedance is too low, useful conversion gain will be sacrificed, as shown in Fig. 4-30.

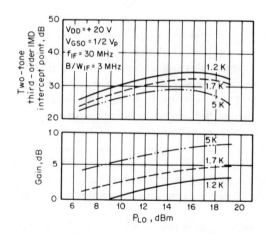

FIGURE 4-30   Gain and two-tone intercept point vs. local oscillator drive.

4-10-6   Mixer Performance

Quad FET arrays with both high and low pinchoff-voltage levels were used in evaluation of the active double-balanced mixer; the mixer exhibited clearly superior characteristics, compared to equivalent small-signal passive double-balanced mixers. The low- to medium-pinchoff-voltage FET array performed slightly better than the high-pinchoff devices solely because of a limitation in available local oscillator power.

4-11   THE COMMUTATION MIXER

Unlike the transconductance FET mixers or the conventional diode-ring mixer, the commutation mixer relies on the switching action of the Siliconix Si8901 quad MOSFETs to effect mixing action. Ideally, we would expect little noise contribution. However, the mixer is not perfect resulting in both conversion loss and a defined noise figure.

This loss and defined noise figure results from two related factors. First, is the r_{DS} of the MOSFETs relative to the signal impedance, R_g, and the intermediate-frequency, i-f, impedance, R_L. Second, and a more common and expected factor, is the loss attributed to signal conversion to undesired frequencies. These include the image and harmonic frequencies.

The effect of r_{DS} of the MOSFETs may be determined from the analysis of the equivalent circuit shown in Fig. 4-31, by first assuming that the local oscillator waveform is an idealized square wave. It is not, of course, but assuming it is simplifies the analysis.

Figure 4-31, showing switches rather than MOSFETs, also identifies the ON-state resistance, r_{DS}, as well as the OFF-state resistance, r_{OFF}. The latter can be disregarded as it is generally extremely high. The ON-state

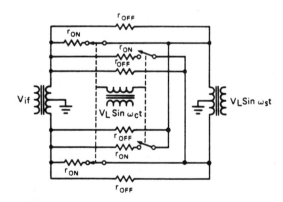

FIGURE 4-31   Equivalent circuit of a commutation mixer. Switches replace FETs. (From Oxner, "Designing FET Balanced Mixers for High Dynamic Range," courtesy of Siliconix Incorporated.)

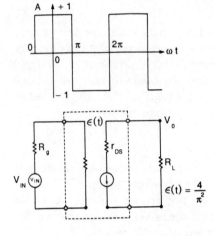

FIGURE 4-32 Derivative equivalent circuit. (From Oxner, "Designing FET Balanced Mixers for High Dynamic Range," courtesy of Siliconix Incorporated.)

$$\epsilon(t) = \frac{4}{\pi^2}$$

resistance, r_{DS}, together with the source and load impedances (i.e., the signal and i-f impedances) directly affects the conversion efficiency and noise figure.

If we assume that the local-oscillator excitation is an idealized square wave, the switching action may be represented by the Fourier series as,

$$f(x) = \frac{1}{2} + \frac{2}{\pi} \sum_{n=1}^{\infty} \frac{\sin(2n-1)\,\omega\,t \cdots}{(2n-1)} \qquad (4\text{-}76)$$

The switching function, $\epsilon\,(t)$, shown in the derivative equivalent circuit of Fig. 4-32, is derived from the magnitude of this Fourier series expansion as a *power* function by squaring the first term of Eq. (4-76), i.e., $(2/\pi)^2$.

The available power that can be delivered from a generator of rms open-circuit terminal voltage, V_{in}, and internal resistance, R_g, is

$$P_{\text{av}} = \frac{V_{\text{in}}^2}{4\,R_g} \qquad (4\text{-}77)$$

or, in terms shown in Fig. 4-33,

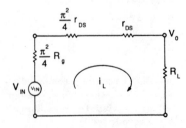

FIGURE 4-33 The power-loop circuit with all elements equivalent. Based on the transfer function $\epsilon(t) = 4/\pi^2$. (From Oxner, "Designing FET Balanced Mixers for High Dynamic Range," courtesy of Siliconix Incorporated.)

$$P_{av} = \frac{V_{in}^2}{\pi^2 R_g} \tag{4-78}$$

the output power, deliverable to the intermediate-frequency port, is

$$P_{out} = \frac{V_o^2}{R_L} \tag{4-79}$$

To arrive at V_o, we first need to obtain the loop current, i_L, which from Fig. 4-33 offers,

$$i_L = \frac{V_{in}}{\dfrac{\pi^2}{4}(R_g + r_{DS}) + R_L + r_{DS}} \tag{4-80}$$

then,

$$V_o = \frac{V_{in}^2 \, R_L}{\dfrac{\pi^2}{4}(R_g + r_{DS}) + R_L + r_{DS}} \tag{4-81}$$

Combining Eqs. (4-79) and (4-81),

$$P_{out} = \frac{V_{in}^2 \, R_L}{[\dfrac{\pi^2}{4}(R_g + r_{DS}) + R_L + r_{DS}]^2} \tag{4-82}$$

Conversion efficiency—a loss—is then calculated from the ratio of P_{av} and P_{out},

$$L_c = 10 \log \frac{P_{av}}{P_{out}} \qquad dB \tag{4-83}$$

Substituting Eq. (4-78) for P_{av} and Eq. (4-82) for P_{out}, we obtain

$$L_c = 10 \log \frac{[\dfrac{\pi^2}{4}(R_g + r_{DS}) + R_L + r_{DS}]^2}{\pi^2 R_L \, R_g} \tag{4-84}$$

The conversion loss represented by Eq. (4-84) is for a narrowband double-balanced mixer with image and sum frequency (RF + LO) ports shorted. If we let $r_{DS} = 0$ and resistively terminate both ports, the minimum attainable conversion loss reduces to

$$L_c = 10 \log \frac{4}{\pi^2} \qquad dB \tag{4-85}$$

which reduces to $L_c = -3.92$ dB. In a practical sense we must add 3.92 dB to the results of Eq. (4-84).

Equation (4-84) can be examined for various values of source (RF) and load (i-f) impedances as well as r_{DS} by graphical representation, shown in Fig. 4-34, remembering that a nominal 3.92 dB must be added to the values obtained.

4-11-1   Intermodulation Distortion in the Commutation Mixer

Equation (4-84) and Fig. 4-34 represent performance based on a square-wave local oscillator excitation. Optimum intermodulation distortion demands that the excitation exhibit a duty cycle of exactly 50%. For square-wave excitation this poses no great problem. Because of the sinusoidal local-oscillator waveform commonly used there are some inherent causes for intermodulation distortion. These result from the nonlinear transfer characteristic of the MOSFETs, shown in Fig. 4-35. Because of the threshold voltage inherent in the enhancement-mode MOSFETs used in the commutation mixer, to realize the necessary 50% duty cycle requires some form of offset bias.

Walker[7] has derived an expression showing the predicted improvement in the relative level of two-tone third-order intermodulation products (IMD$_3$) as a function of the rise and fall times of the local-oscillator waveform.

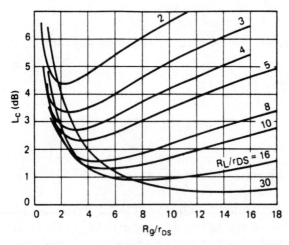

FIGURE 4-34 Insertion loss as a function of r_{DS}, R_L, and R_g. (From Oxner, "Designing FET Balanced Mixers for High Dynamic Range," courtesy of Siliconix Incorporated.)

[7]Walker, H.P., "Sources of Intermodulation in Diode-Ring Mixers," *The Radio and Electronic Engineer,* **46**(5), pp. 247–255, May 1967.

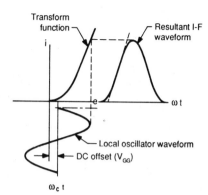

FIGURE 4-35 Effect of sinusoidal LO waveform on IF linearity. (From Oxner, "Designing FET Balanced Mixers for High Dynamic Range," courtesy of Siliconix Incorporated.)

$$\text{IMD} = 20 \log \frac{[t_r \, \omega_{\text{LO}} \, \dfrac{V_s}{V_c}]^2}{8} \qquad (4\text{-}86)$$

where V_c = peak-to-peak local-oscillator voltage

V_s = peak signal voltage

t_r = rise and fall time of V_c

ω_{LO} = local-oscillator frequency.

Equation (4-86) offers several interesting aspects pertaining to performance. (1) Since by lowering V_s the IMD is improved, if R_g is reduced, performance improves! (2) The higher the local-oscillator voltage, the better. (3) By providing a square-wave local-oscillator waveform we achieve an infinite improvement in IMD performance!

Equation (4-86) clearly identifies the necessity of either a high local-oscillator voltage or square-wave excitation. At low frequencies a square-wave source may be practical, but for high-frequency operation another means may be necessary.

A resonant tank offers an easy source for high voltages. These voltages may be calculated

$$V = (P \, Q \, X)^{1/2} \qquad (4\text{-}87)$$

where P = power delivered to the resonant tank circuit,

Q = *loaded Q* of the tank circuit, and

X = reactance of the Si8901 gate-to-gate capacity.

The effect of resonant gate voltage on performance for a prototype commutation mixer using the Si8901 is shown in Fig. 4-36. The two-tone, third-order intercept point of the Si8901 mixer, shown schematically in Fig. 4-37 is provided in Fig. 4-38. At a signal frequency of 100 MHz and an *i-f* of 30 MHz the overall single-sideband noise figure was 8 dB. The 2-dB compression level and desensitization level was +30 dBm!

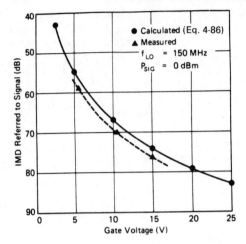

FIGURE 4-36   Effect of gate voltage on IMD
performance. (From Oxner, "Designing FET
Balanced Mixers for High Dynamic Range,"
courtesy of Siliconix Incorporated.)

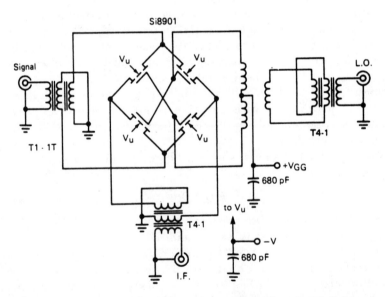

FIGURE 4-37   Schematic of prototype commutation double-balanced
mixer. (From Oxner, "Designing FET Balanced Mixers for High Dynamic
Range," courtesy of Siliconix Incorporated.)

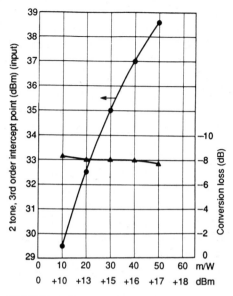

FIGURE 4-38 Intercept point and conversion loss of commutation double-balanced mixer. (From Oxner, "Designing FET Balanced Mixers for High Dynamic Range," courtesy of Siliconix Incorporated.)

4-12 OSCILLATORS

FETs have some fundamental advantages over bipolar transistors in oscillators. An outstanding advantage is their very low sideband noise. In addition, they may be designed so that the frequency of oscillation is relatively independent of bias current. This reduces the effects of drift during temperature fluctuations. The parasitic capacitances of FETs are known to be voltage-dependent, and this can be turned to great advantage since it makes possible easily designed VTOs (voltage-tuned oscillators).

Although noise is often difficult to characterize because of its random or nondeterministic nature, it is possible to differentiate various forms of noise through an understanding of the gaussian distribution of noise about an RF carrier. The three major forms of noise are (1) low-frequency noise ($1/f$), (2) thermal noise ($4kTBR$), and (3) shot noise ($\bar{i_n}$). These types of noise can be identified from their relationship to the main RF carrier. For example, low-frequency noise predominates very close to the carrier and falls to insignificant levels when it is displaced more than 250 Hz from the carrier. Thermal noise plays the dominant role in the midfrequency region from the $1/f$ noise decay point to approximately 20 kHz from the carrier. It is commonly associated with equivalent resistance where the rms value of noise voltage of the Thevenin generator becomes the classic $\sqrt{4kTBR}$. Noise appearing beyond 20 kHz is known as shot noise, and is directly attributed to noise current. Because of the typically uniform distribution of shot noise, it is often referred to as "white noise."

Generally oscillators are represented as narrowband circuits; consequently most interest in noise centers about a reasonably narrow band. The voltage that results when white noise is passed through any relatively narrow bandpass may be described by

$$v(t) = x_1(t)\cos \omega_{ot} + x_2(t)\sin \omega_{ot} \qquad (4\text{-}88)$$

where ω_o is the midband angular frequency and $x_1(t)$ and $x_2(t)$ are uncorrelated functions of time which vary slowly and randomly about zero in a manner described by the gaussian probability density.

4-12-1 Origins of Oscillator AM Noise

Although an oscillator tends to produce a wave that is nearly sinusoidal, there are other fluctuations present. When the energy in the frequency domain close to the carrier is observed on a spectrum analyzer, noise appears as a modulation phenomenon. This observation would be greatly enhanced if the noise contribution were coherent and consisted of discrete sideband frequencies. The major component of AM noise is low-frequency noise $(1/f)$. Both thermal and shot noise are relatively insignificant segments of AM noise compared with $1/f$. A graph of AM noise versus frequency removed from the carrier is shown in Fig. 4-39.

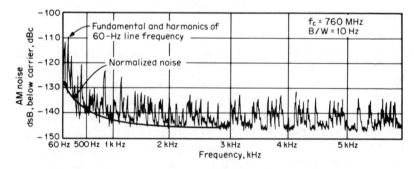

FIGURE 4-39 AM noise vs. frequency removed from the carrier.

4-12-2 Design of a VHF Oscillator

Important design considerations for best oscillator performance include using a FET with high forward transconductance (g_{fs}), maintaining the gate at ground potential, and keeping a high unloaded tank Q. The g_{fs} reduces the effective noise resistance. The grounded gate reduces the noise voltage resulting from the product of the gate leakage current and the series gate resistance. The high tank-circuit Q serves as an effective filter for the sideband noise energy.

The example used to illustrate an ultra-low-noise design is somewhat extraordinary for a circuit employing a FET. The FET chosen was the Siliconix U310, which has a forward transconductance value higher than 18 mmhos at zero bias $(V_{GS} = 0)$. The oscillator basically consists of two coaxial resonators, one for the FET source and the other for the drain. Oscillation is established by capacity coupling between the two resonators; output coupling is derived from the magnetic coupling which exists at the open ends of the resonators. Optimum resonator Q is achieved by designing the coaxial resonators for a characteristic impedance of 75 Ω. The oscillator circuit is shown in Fig. 4-40.

The technique to establish the proper resonator length for the desired frequency requires a first-order approximation of the anticipated capacitive fringing which derives from both the FET and the feedback network. A short-circuited coaxial transmission line is theoretically resonant at a quarter-wavelength of the resonating frequency, except for the effects of fringe field capacitance. At resonance,

$$X_L = X_C \qquad (4\text{-}89)$$

If the fringe capacitance is known, X_C can be calculated as

$$X_C = \frac{1}{\omega C} \qquad (4\text{-}90)$$

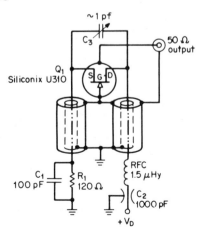

FIGURE 4-40 Oscillator circuit.

From this, the resonator length can be determined as

$$X_C = \tan \beta l \tag{4-91}$$

In making these calculations, a Smith chart is invaluable, as is shown in the following illustration:

Frequency of oscillation	$= 760$ MHz
FET b_{igs} (from data sheet)	$= 16$ mmhos
Capacitance from b_{igs}:	$C_{gs} = 3.4$ pF
Allow for stray capacitance	
and the feedback network:	$C_s = \underline{1.5 \text{ pF}}$
	4.9 pF

Thus $X_C = j0.57$ (normalized to 75 Ω)

Locate 0.57 on the Smith chart. The wavelength toward the load = 0.081λ. Since a wavelength at 760 MHz is 39.5 cm, the resonator cavity length is simply

$$39.5 \times 0.081 = 3.20 \text{ cm (1.26 in)} \tag{4-92}$$

In the completed FET coaxial oscillator circuit, the output-coupling loop consists of a single turn made fast to the cavity by the BNC flange and the FET itself.

4-12-3   Conclusions

Measured performance of the oscillator is shown in Table 4-7A; AM noise measurements in a 10-Hz bandwidth are shown in Table 4-7B.

TABLE 4-7A   OSCILLATOR MEASURED PERFORMANCE AT 25°C

V_{DD} (V)	+10	+15	+20	+25
I_D (mA)	15	16.2	18.2	21
P_{out} (dBm)	+6.6	+15.2	+18.3	+20
Frequency (MHz)	725	742.7	754.7	762.9

TABLE 4-7B   AM NOISE MEASUREMENT

Frequency displaced from carrier	dBc
50 Hz	−130
500 Hz	−139
1 kHz	−143.5
5 kHz	−146

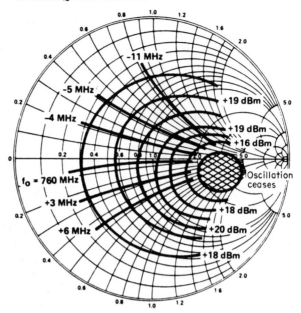

FIGURE 4-41 Reike diagram.

The Reike diagram shown in Figure 4-41 makes possible the accurate prediction of expected power output and operating frequency and the oscillator feeding directly into a mismatched load. Expansion of the Reike diagram to show frequency versus transmission-line length (in degrees) will allow prediction of the long-line effect on oscillator stability.

4-13 NOISE TEMPERATURE AND NOISE FIGURE

Johnson noise, defined as a nonperiodic ac voltage fluctuation, places a limit on the sensitivity of amplifiers and mixers. All high-frequency transistors, from audio frequencies up, whether FETs or bipolars, as well as all passive elements, reach a well-defined limit of performance based on Johnson noise.

Historically, active transistors have been categorized by their *noise figure;* the lower the value, the better the device. Perhaps two decades ago, hidden within the confining technology of ultra-low-noise parametric amplifiers and hydrogen masers, was a unique term of performance called noise temperature.

Since Johnson noise is an effect of electron agitation, it is reasonable to assume that at absolute zero ($-273.18°C$) Brownian movement, and hence electron agitation, ceases. Consequently, since our standard operating temperature is generally regarded as $20°C$, the standard noise temperature T_o of a passive device, such as a resistor, would be 293 K. By universal scientific consent, and for convenience in mathematical analysis, T_o has been established at 290 K.

Again, since Johnson noise is electron agitation, it is temperature-dependent. However, it must be understood and appreciated that active components, such as transistors, can exhibit noise temperatures *differing* from their operating temperatures as well as their ambient temperatures! This exhibited noise temperature is the "effective input noise temperature" T_e of the active device.

Many readers may be aware of the term "input available noise power" as $P = kTB$, where k is Boltzman's constant, B is bandwidth in hertz, and T is temperature. The noise output of an active transistor is the sum of the input noise and the noise contributed by the device:

$$N_{\text{power}} = GkB(T_{\text{in}} + T_e) \tag{4-93}$$

where G is the gain and T_e is that term defined above as effective input noise temperature.

The definition of "noise figure" is the ratio of the total noise power delivered when the noise temperature at its input is 290 K to that of the input. In other words, by formula:

$$NF = \frac{N_{\text{power}}}{GkB(290)} \tag{4-94}$$

Combining Eqs. (4-93) and (4-94), we obtain

$$NF = 1 + \frac{T_e}{290} \tag{4-95}$$

Since NF is generally identified in decibels, the above formula (4-95) may be expanded to

$$NF_{\text{dB}} = 10 \log \left(1 + \frac{T_e}{290}\right) \tag{4-96}$$

or

$$T_e = 290 \left[\left(\text{antilog} \frac{NF}{10}\right) - 1\right] \quad K \tag{4-97}$$

For example, the U310/NZA has a noise figure of 2.7 dB at 450 MHz. What is its noise temperature?

$$T_e = 250 \text{ K}$$

4-13-1 Optimum Noise Matching of Amplifiers

In any amplifier or linear two-port, whether performing at audio or at some high frequency, any noise generated by the input stage tends to obscure or mask the amplified signal appearing at the output. This is so because noise is as readily amplified as signal. Consequently, it should be a designer's goal to minimize this input-contributed noise. To do so will greatly enhance the small-signal performance of the amplifier. Such an amplifier is said to be "noise matched."

Noise is generated by thermal agitation within resistive elements. Pure reactive elements (such as L and C) do not contribute noise. Consequently, when the total input admittance is a pure conductance, the noise figure of the amplifier is minimized—that is, when

$$j y_{in} = 0$$

From this, quite apart from bandwidth requirements, it is desirable to have the input circuit resonant at band center. Any losses within the input stage will increase the noise figure even if these losses have an effective noise temperature of zero!

It is known that the equivalent input noise resistance for JFETs is

$$R_{eq} = \frac{0.67}{g_{fs}} \tag{4-98}$$

It has been established that to obtain the lowest first-stage noise figure, the input admittance is a root function of the ratio of a complex transistor conductance G_b to its equivalent noise resistance R_{eq}; thus

$$G_{s(opt)} = \sqrt{\frac{G_b}{R_{eq}}} \tag{4-99}$$

where G_b is the additive combination of both the shunt input conductance as a function of FET gate current I_g and the equivalent conductance of the gate-to-source as a function of frequency.

The best way to determine optimum noise source resistance for most amplifiers is to measure it! This is not always an easy task. Generally the optimum source resistance turns out not to be a resistance *per se*, but an optimum source admittance, that is, a complex quantity consisting of both conductance and susceptance, viz., $Y_{s(opt)} = G + jB$.

Optimum source admittance for FETs varies both with frequency and with configuration; common-source input stages generally have lower values of admittance than would the same FET in a common-gate configuration. Likewise, since gate current is critical to the optimum input noise admittance, the FET geometry plays an important role.

For the Siliconix U310 (NZA geometry), the optimum input source admittance for the *common-gate configuration* has been determined:

450 MHz $R_{eq} = 166\ \Omega$ in parallel with $X_l = 120\ \Omega$

4-13-2   Optimum Noise Bandwidth of Amplifiers

The bandwidth of an amplifier contributes to its faithfulness in reproduction as well as to its dynamic range. This does not mean that the wider the bandwidth, the better the fidelity, for such is not always the case. The bandwidth must be sufficiently broad to accept the total signal, yet not so wide that spurious and Johnson noise mask the weaker signals.

Dynamic range simply means that the amplifier not only detects the *weakest* possible signal but also will not readily overload under strong input signals. A steady-state amplifier has far different design requirements than a transient (fast-signal) amplifier. For the latter, dynamic range also suggests purity of response.

The detectability of weak signals involves many aspects; what will be addressed here is simply the noise bandwidth.

A narrowband amplifier will amplify only a narrow spectrum of signals. If the passband is less than the signal spectrum, the fidelity will be impaired because the output signal will reach only a fraction of its peak value. Since noise power is a linear function of bandwidth, a narrowband amplifier will not have the noise susceptibility of its wide-bandwidth counterpart. Even though the noise is restricted because of bandwidth, signal fidelity is also restricted *if* the signal spectrum exceeds the amplifier bandwidth.

As the passband is widened, the signals are reproduced more faithfully until the output reaches its peak value. Simultaneously, the noise power increases. It should be evident that eventually an optimum bandwidth occurs at which there is no further improvement in signal fidelity. Any additional increases in bandwidth only generate increasingly greater amounts of noise power that do nothing beneficial to the signal. In fact, on the contrary, small signals are masked, thus reducing the dynamic range of the amplifier.

An amplifier with too narrow a passband restricts the fidelity (as any audiophile is well aware), and too wide a passband contributes excessive noise, reducing the signal-to-noise ratio of the amplifier.

Even the "ideal" noiseless amplifier must always be coupled to a resistive source, which, at an effective noise temperature of 290 K, will thus offer a minimum detectable threshold of approximately −204 dBW/Hz. In the practical sense the amplifier will contribute noise which

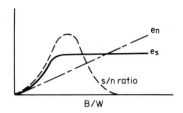

FIGURE 4-42   Effect of optimum bandwidth on signal to noise performance of an amplifier.

is, among other things, proportional to its bandwidth, resulting in the appearance at the output of an apparent threshold noise level whose magnitude is far greater than the ideal noiseless amplifier and perhaps even greater than the signal!

Previously we pointed out the importance of noise-matching the amplifier to offer optimum performance. However, an optimally matched input, by definition, forced narrow-bandwidth performance. Consequently, it appears that optimum bandwidth/optimum noise is a tradeoff design problem. Figure 4-42 represents how bandwidth affects signal/noise performance of the amplifier.

BIBLIOGRAPHY

Ghausi, M. S.: *Principles and Design of Linear Active Circuits,* McGraw-Hill, New York, 1965.

Hetterscheid, W. Th. H.: *Transistor Bandpass Amplifiers,* Philips Technical Library, Eindhoven, the Netherlands, 1964.

Klaus, Herbert L., Charles W. Bostian, and Frederick H. Raab: *Solid State Radio Engineering,* Wiley, New York, 1980.

Linvill and Gibbons: *Transistors and Active Circuits,* McGraw-Hill, New York, 1961.

Motorola Semiconductor Products Division, Phoenix, Ariz.:
   Motorola Appl. Notes AN-423: "FET RF Amplifier Design Techniques."

   Motorola Appl. Notes AN-215: "RF Small Signal Design Using 2-Port Parameters."

   Motorola Appl. Notes AN-166: "Using Linvill Techniques for RF Amplifiers."

Raab, F. H.: "The Class BD High-Efficiency RF Power Amplifier," *IEEE J. Solid-State Circuits,* SC-**12**(3), June 1977.

————: "High Efficiency RF Power Amplifiers," *Ham Radio,* **7**, October 1974.

Valley, George, and Henry Wallman: "Vacuum Tube Amplifiers," Radiation Laboratory Series, reprint by Dover Publications, New York, 1965.

5

ANALOG SWITCHES

5-1 THE FET AS AN ANALOG SWITCH

The field-effect transistor, in the ON condition, contains a conducting channel of either n-type or p-type carriers. The carriers, in traversing from source to drain or drain to source, do not cross p-n junctions of the type encountered in bipolar transistor or diode switches; thus there is no inherent offset voltage in the ON switch. Signal currents can typically pass equally well in either direction in the ON switch and are blocked equally well in either direction by the OFF switch. This chapter presents some of the important characteristics of the FET as an analog switch and shows methods of driving the switch (controlling the ON-OFF status).

5-2 DC EQUIVALENT CIRCUITS

Some of the basic switching modes to be considered are shown in Fig. 5-1. The "voltage-mode" switch (a) and the "current-mode" switch (b), for example, may be used for multiplexing many signals into a common

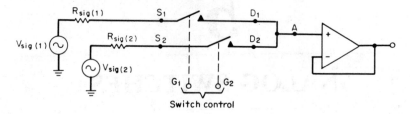

(a) Switch with high-impedance load

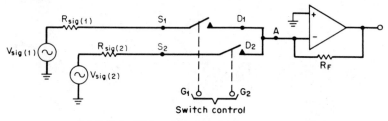

(b) Switch with low-impedance load

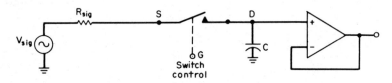

(c) Switch used in sample and hold system

FIGURE 5-1   Analog switch circuits: *(a)* switch with high-impedance load; *(b)* switch with low-impedance load; and *(c)* switch used in sample-and-hold system.

point such as the amplifier inputs shown. The Fig. 5-1*b* circuit is often used for the selective summing of two or more signals into a low-impedance summing node and in digital-to-analog converters. The Fig. 5-1*c*-type sample-and-hold circuit finds applications in analog-to-digital converters. For these circuits we will show how various FET types of the "family tree" may be used. We make the initial assumption that in the ON state the value of V_{DS} is small and may be of either polarity. For the FET types considered, the value of channel conductance is approximately a linear function of gate-to-source voltage, using as a reference the gate-source cutoff $V_{GS(off)}$ or threshold voltage $V_{GS(th)}$. Figure 5-2 shows this characteristic. Also shown in Fig. 5-2 is the effect that the body-source voltage has upon channel conductance of the MOS devices. The body, in effect, functions as a "back" gate, and its effect upon channel conductance must be considered.

The perfect switch would have infinite resistance (zero conductance) when open and zero resistance (infinite conductance) when closed. While the FET is not a perfect switch, there are many applications in which this deviation from perfection is unimportant. This statement can be justified by an analysis of the implications of the circuits shown in Fig. 5-3. The general two-port network in Fig. 5-3a couples the signal source V_{SIG} to a resistive load R_L. The network can be characterized by its terminal voltages and currents, V_1, V_2, I_1, and I_2. Figure 5-3b shows the equivalent circuit of a FET switch in the OFF state. In this condition, the "source" and "drain" are not connected to one another; however, two leakage current sources, I_S and I_D, are present. The same device is shown in the ON state in Fig. 5-3c. The FET gate is connected to its source. The leakage I_0 is that of the driver. The following typical values are assumed for the circuit. The switch characteristics are those of the widely used analog switch type DG181AP.

$$V_{SIG} = \pm 10 \text{ V} \qquad r_{DS} = 30 \text{ }\Omega$$
$$R_{SIG} = 10 \text{ }\Omega \qquad I_S = I_D = 1 \text{ nA}$$
$$R_L = 200 \text{ k}\Omega \qquad I_0 = 2 \text{ nA}$$

In the following calculations the effect of leakage current is expressed in terms of error percentage.

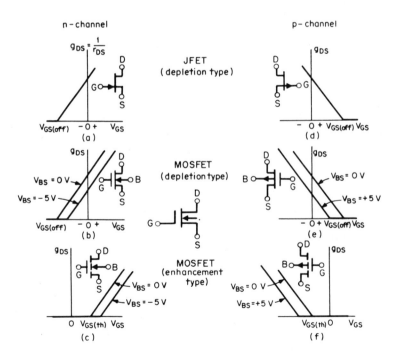

FIGURE 5-2 Channel conductance vs. gate-source voltage.

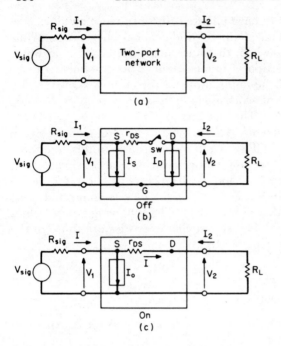

FIGURE 5-3 DC equivalent circuits.

5-2-1 OFF Condition Calculation (Fig. 5-3b)

$$I_1 = I_S = 1 \text{ nA}$$

$$V_{\text{SIG}} - V_1 = I_1 \cdot R_{\text{SIG}} = (1 \text{ nA})(10 \ \Omega) = 10 \text{ nV} \qquad (5\text{-}1)$$

$$\% \text{ error in } V_1 = \frac{(10^{-8} \text{ V})(10^2)}{10 \text{ V}} = 1 \times 10^{-7}\%$$

$$I_2 = I_D = 1 \text{ nA}$$

$$V_{2(\text{off})} = I_2 R_L = (1 \text{ nA})(200 \text{ k}\Omega) = -200 \ \mu\text{V} \qquad (5\text{-}2)$$

$$\% \text{ error in } V_{2(\text{off})} = \frac{(2 \times 10^{-4})(10^2)}{10} = 0.002\%$$

referred to V_{SIG} (full scale).

5-2-2 ON Condition Calculation (Fig. 5-3c)

$$I_2 = \frac{V_2}{R_L} \cong \frac{V_{\text{SIG}}}{R_L + R_{\text{SIG}} + r_{DS}} = 42 \ \mu\text{A} \qquad (5\text{-}3)$$

$$V_{\text{SIG}} - V_2 \cong (42 \ \mu\text{A})(40 \ \Omega) = 1.7 \text{ mV}$$

$$\% \text{ error in } V_2 = \frac{(1.7 \times 10^{-3})(10^2)}{10} = 1.7 \times 10^{-2} = 0.017\%$$

referred to V_{SIG} (full scale)

The error resulting from I_o is approximately 20 nV or $2 \times 10^{-7}\%$.

The foregoing calculations indicate that for many applications the performance of the FET equivalent circuits in Fig. 5-3 is a good approximation of the perfect switch. In particular, the OFF condition leakage currents contribute only a negligible portion of total error.

5-3   THE JFET AS A SWITCH

A suitable driving circuit must be considered when assessing the performance of the JFET as a switch. Such a circuit is shown in Fig. 5-4.

Note that Q_1 is an n-channel JFET, Q_2 is an enhancement-mode p-channel MOSFET, and Q_3 is an enhancement-mode n-channel MOSFET. A V_{in} of -15 V will turn Q_3 OFF and Q_2 ON, so that S_1 and G_1 are connected. With $V_{\text{G1S1}} = 0$, Q_1 is ON. If V_{GS} of Q_1 were allowed to vary, switch resistance modulation would occur, introducing a source of error. Figure 5-5 shows the equivalent circuit of the ON switch.

The suggested driving circuit of Fig. 5-4 avoids r_{DS} modulation at low frequencies. Typically the positive supply voltage is $+15$ V and the negative supply voltage is -15 V. In order for V_{GS} to change, current must flow through Q_2, which is ON. There are only two possible current paths through Q_2: through Q_3, which is OFF and subject only to variations in leakage current, or into the gate of Q_2, which is also subject to leakage current. Since both paths through Q_2 provide only negligible changes in V_{GS}, their effect in the circuit may be ignored. As the analog frequency

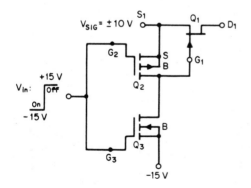

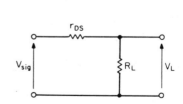

FIGURE 5-4   JFET switch control circuit.

FIGURE   5-5   Error   due   to switch ON resistance r_{DS}.

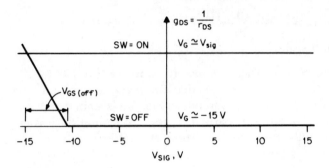

FIGURE 5-6   JFET channel conductance g_{ds} vs. signal voltage.

is increased, capacitance will provide lower-impedance paths, so that some degree of Δr_{DS} is possible at high frequency.

With V_{in} at $+15$ V, Q_2 is OFF and Q_3 is ON; thus G_1 is at -15 V. Q_1 will remain OFF if $V_{SIG} > (V_{G1} - V_{GS(off)})$. $V_{GS(off)}$ is a "negative" voltage for an n-channel FET; thus the negative analog signal is limited by the $V_{GS(off)}$ of Q_1 and the negative supply voltage. The ON and OFF conditions are shown in Fig. 5-6. g_{DS} is constant because with $V_{G1} = V_{SIG}$ imposed by the switch control circuit, $V_{GS} = 0$.

The positive swing of the ON switch is limited only by the maximum voltage ratings of Q_2, and Q_3. In the OFF state the positive swing of the source and drain of Q_1 is limited by the voltage breakdown ratings of both Q_1 and Q_2.

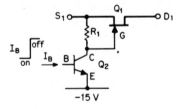

FIGURE 5-7   JFET switch control circuit.

Another driver circuit for the JFET is shown in Fig. 5-7. In the switch ON state, Q_2 is turned OFF; thus the gate of Q_1 is free to follow the analog signal at its source, *provided* the reactance of the gate-to-ground capacitance of Q_1 plus the output capacitance of Q_2 is large compared to the resistance of R_1. At high analog signal frequencies this circuit may develop problems because of the gate's inability to follow the source,

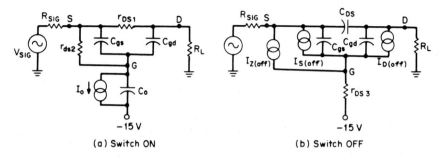

FIGURE 5-8 Equivalent circuits using switch shown in Fig. 5-4.

which results in a modulation of the channel resistance by the signal voltage. This becomes more serious as the analog amplitude is increased.

In the switch OFF state, the gate of Q_1 is clamped to -15 V, and R_1 appears as a load on the analog signal source.

The ON and OFF equivalent circuits for the circuits of Figs. 5-4 and 5-7 are shown in Figs. 5-8 and 5-9. To minimize analog signal feedthrough of the OFF switch, r_{DS3}, r_{ce2}, C_{ds}, C_{gs}, and C_{gd} should be low. In the ON state, the value of r_{DS1} should be low to minimize insertion loss. Minimizing r_{DS1} must be compromised with keeping C_{gs} and C_{gd} low, because designing the FET for lower r_{DS} will usually increase its gate capacitance or increase $V_{GS(off)}$.

5-4 SWITCHING HIGH-FREQUENCY SIGNALS

As a signal frequency is increased, a decrease in OFF isolation rather than degradation of ON performance is the limiting factor. Three factors affect the quality of OFF isolation: selection of the appropriate analog switch, the magnitude of load resistance, and the amount of stray capacitance present in the circuit. The selection of the appropriate

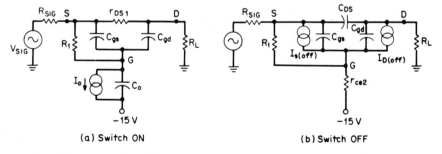

FIGURE 5-9 Equivalent circuits using switch shown in Fig. 5-7.

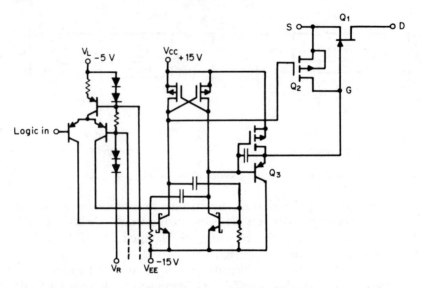

FIGURE 5-10   Schematic diagram of one channel of DG181 analog switch.

analog switch is probably the most important factor. As a rule, product specifications do not provide OFF isolation data in sufficient depth to make switch selection easy. Three steps may be taken to overcome this lack of data: measurement of actual isolation performance, analysis of equivalent circuits to predict isolation, and simplification of the analysis to produce a usable set of design aids.

To illustrate performance analysis, a JFET switch with an integrated drive will be used for switching wideband RF signals. The JFET used as the switch is similar to the 2N3971 (except that $V_{GS(off)}$ is lower).

It is packaged with an integrated circuit driver which permits control with low-level logic signals such as the output from TTL digital circuits. A combined switch pair with driver has the commercial part number DG181BA. Figure 5-10 shows a schematic diagram of one channel of this dual-channel switch. The driver output is similar to the control circuit shown in Fig. 5-4. Replacing the MOSFET Q_3 of Fig. 5-4 with the p-n-p transistor shown in Fig. 5-10 improves OFF isolation because of the lower ON impedance of the p-n-p transistor.

5-4-1   OFF Isolation

The isolation performance test setup and a plot of data are shown in Fig. 5-11. In the equivalent OFF circuit shown in Fig. 5-12, C_o and R_o are output parameters of the driver p-n-p transistor Q_3 (Fig. 5-10). The 4.5-pF source-to-gate capacitance includes the source-to-drain capacitance of Q_2. The improvement in performance that could be obtained

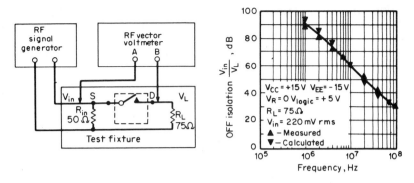

FIGURE 5-11 DG181 OFF isolation test circuit and data.

if R_0 were reduced to zero is shown in the table included in Fig. 5-12. A large part of the feedthrough is via C_{gs} and C_{gd} unless the driver output impedance is low.

In Fig. 5-12, two separate paths exist between source (S) and drain (D): through C_{ds} and through the gate circuit C_{gs} and C_{gd} on the way to the drain. To simplify analysis, it is assumed that I_F (current through C_{ds}) and I_T (the current flowing out of the tee network) are independent

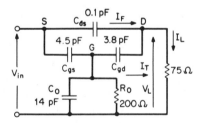

In the test fixtures, inputs were shielded from outputs, and RF decoupling was provided on all dc connections. Great care was taken in the mechanical layout of fixtures, to minimize stray capacitance. The characteristic impedance of video transmission lines, 75 Ω, was issued as the value of load resistance. Voltage measurements were made with an RF vector voltmeter, H/P Model 8405A.

COMPARISON OF IDEAL DRIVER CASE ($I_T = 0$) WITH ACTUAL PERFORMANCE OF DG181 ANALOG SWITCH

R_0	I_F	I_T	I_L	$\|V_{in}/V_L\|$ (dB)
0	1.41 $\underline{/90°}$ μA	0	1.41 $\underline{/90°}$ μA	66.5
200	1.41 $\underline{/90°}$ μA	2.91 $\underline{/163°}$ μA	3.58 $\underline{/141°}$ μA	58.4

FIGURE 5-12 Equivalent circuit of DG181 with 75-Ω load.

of one another. While this assumption is not entirely valid, if OFF isolation is greater than 20 dB, it yields excellent agreement with measured results.

The transfer functions for I_F and I_T are

$$I_F = \frac{j\omega C_{DS} R_L}{1 + j\omega C_{DS} R_L} \tag{5-4}$$

and

$$I_T = \frac{j\omega C_{GS} V_{\text{in}}}{1 + R_L/R_0 + (C_{GS} + C_0)/C_{GD} + jR_L/R_0 \left[\omega R_0(C_{GS} + C_0 - 1/\omega R_L C_{GD}\right]} \tag{5-5}$$

R_0 and C_0 are the output resistance and capacitance of the FET driver circuit. Equation (5-5) shows that as $R_0 \to 0$, $I_T \to 0$. It is thus possible to reduce R_0 and make I_T an arbitrarily small value; however, I_F remains to be dealt with. I_F is the sum of the currents through C_{DS} (device capacitance) and C_{stray} (additional wiring capacitance, etc.). It may be the dominant current at certain frequencies. Table 5-1 shows that for the DG181 switch I_F is dominant at 1 MHz and I_T is dominant at 100 MHz. The separate expressions derived for I_F and I_T make it relatively simple to evaluate the effect of varying certain parameters to minimize I_L (maximize isolation).

Several multiple-switch configurations may be used to achieve an impressive increase in OFF isolation under otherwise difficult conditions. Probably the most effective multiple-switch configuration is the TEE, which is shown in Fig. 5-13. In the TEE, S_2 operates out of phase with S_1 and S_3, and provides two stages of isolation. The input to S_3 is the isolation leakage of S_1 working into an $R_L = r_{DS2}$. This multiple-switch arrangement brings about a considerable improvement in OFF isolation, but at the expense of doubling switch ON resistance. This increases the ON insertion loss.

TABLE 5-1 VARIATIONS IN CURRENT WITH FREQUENCY FOR DG181

f (MHz)	I_F	I_T	I_L	$\|V_{in}/V_L\|$ (dB)	$\|C_{eq}\|$ (pF)
1.0	141 $\underline{/90°}$ nA	30.3 $\underline{/178°}$ nA	145 $\underline{/102°}$ nA	+86	0.103
4.0	563 $\underline{/90°}$ nA	481 $\underline{/173°}$ nA	784 $\underline{/128°}$ nA	+72	0.139
10.0	1.41 $\underline{/90°}$ μA	2.91 $\underline{/163°}$ μA	3.58 $\underline{/141°}$ μA	+58	0.254
40.0	5.63 $\underline{/90°}$ μA	31.9 $\underline{/128°}$ μA	36.5 $\underline{/123°}$ μA	+38	0.648
100.0	14.1 $\underline{/90°}$ μA	99.6 $\underline{/101°}$ μA	113 $\underline{/100°}$ μA	+28	0.803

$V_L = 224$ mW; $R_L = 75$ Ω; $C_{DS} = 0.1$ pF; $C_{\text{stray}} = 0$.

NOTE: The equivalent circuit shown in Fig. 5-12 was used to calculate the results shown in Table 5-1.

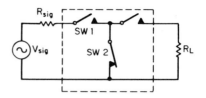

FIGURE 5-13 TEE for isolation improvement.

5-4-2 ON Attenuation

The ON equivalent circuit of a DG181 switch is shown in Fig. 5-14. ON performance is essentially independent of frequency for any load capacitance likely to be used. The ON resistance $r_{DS(on)}$ causes an insertion loss which is basically constant; phase shift is negligible.

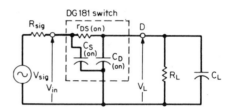

FIGURE 5-14 DG181 ON equivalent circuit.

With the test fixture described in Fig. 5-11, measured ON performance was observed as shown in Fig. 5-15.

The transfer function for the ON switch is

$$\frac{V_L}{V_{in}} = \frac{R_L/(R_L + r_{DS(on)})}{1 + jf\{2\pi[(R_L r_{DS(on)})/(R_L + r_{DS(on)})]\,[C_{D(on)} + C_L]\}} \tag{5-6}$$

$$f_o = \frac{1}{2\pi\left(\dfrac{R_L\,r_{DS(on)}}{R_L + r_{DS(on)}}\right)\left(C_{D(on)} + C_L\right)}$$

FIGURE 5-15 DG181 ON performance.

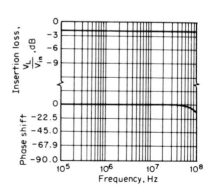

Insertion loss is computed from the numerator as (neglecting C_D and C_L)

$$\text{ON insertion loss, dB} = 20 \log \frac{R_L}{R_L + r_{DS(on)}} \qquad (5\text{-}7)$$

The insertion loss, OFF isolation, and ON/OFF ratio over a range of load resistance are

Switch	$r_{DS(on)}$ (Ω)	R_L (Ω)	Insertion loss (dB)	OFF isolation (dB) $f = 10$ MHz $C_{stray} = 0.1$ pF	ON/OFF (dB)
DG181	25	100	2.0	55.9	53.9
		75	2.5	58.4	55.9
		50	3.6	61.9	58.3

The frequency response of the ON transfer function has the normalized form shown in Fig. 5-16.

5-5 THE MOSFET SWITCH

The control circuit for the MOSFET switch can be simpler than for the JFET because its gate may be positive or negative with respect to source and drain. Typically the gate is switched between two fixed voltages, the sum of which must exceed the peak-to-peak analog voltage by at least the magnitude of $V_{GS(off)}$ or $V_{GS(th)}$ plus enough additional

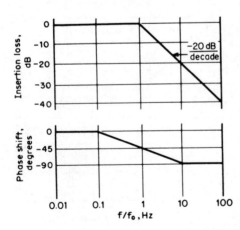

FIGURE 5-16   ON frequency response.

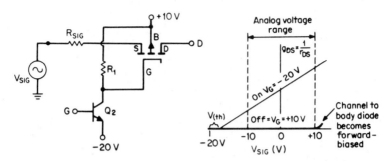

FIGURE 5-17 PMOS channel conductance g_{ds} vs. signal voltage.

V_{GS} bias to produce an adequate g_{ds}. As an example, consider the *p*-channel enhancement-mode MOSFET switch and driver shown in Fig. 5-17. The gate is connected to +10 V (through R_1) for the switch OFF state and to −20 V (via Q_2) for the switch ON state. In the ON state, the channel conduction g_{ds} is a function of the signal voltage as indicated because both gate and body are connected to fixed voltages (−20 V and +10 V). If the source or drain is allowed to swing more positive than +10 V, they become forward-biased with respect to the body, a condition which should typically be avoided. If source and drain approach −20 V, channel conductance g_{ds} will approach zero.

5-6 THE CMOS SWITCH

As noted in Fig. 5-17, the typical PMOS or NMOS switch circuit will exhibit a variation in ON conductance as the analog voltage is varied. This undesirable characteristic can be overcome by paralleling *p*- and *n*-channel MOSFETs, as shown in Fig. 5-18a. For the ON state, the *n*-channel gate is forced positive and the *p*-channel gate is forced negative. Figure 5-18b shows the combined conductance of the two FET switches. The integrated combination of *n*-channel and *p*-channel devices on a common substrate is referred to as complementary MOS (CMOS).

The OFF condition for the CMOS device will be maintained so long as the channel-to-body diodes do not become forward-biased, as shown in Fig. 5-18c.

The major advantages the CMOS construction technique makes to analog switching are:

○ Lower r_{DS} variation than with either P or NMOS over wide analog signal voltage excursions (similar to the performance of a junction FET).

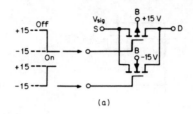

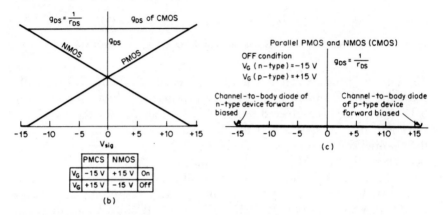

	PMCS	NMOS	
V_G	−15 V	+15 V	On
V_G	+15 V	−15 V	Off

(b)

FIGURE 5-18   Characteristics of CMOS devices.

○ Analog signal range extends to + and − supply voltages. For instance, using the same ±15-V supplies typical of operational amplifiers, the signal-handling capability of the system is limited by the op amp, *not by the switch.*

Figure 5-19 gives a comparison of the characteristic of r_{DS} versus V_{SIG} for typical JFET, PMOS, and CMOS switches. The three characteristic

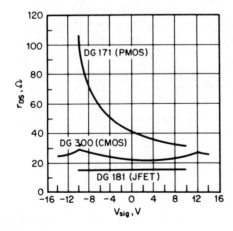

FIGURE 5-19   Performance comparison of three FET switches.

curves shown in Fig. 5-19 were obtained using integrated-circuit analog switches types DG181, DG171, and DG300, respectively, for the JFET, PMOS, and CMOS types.

5-7 THE DMOSFET SWITCH

For those applications where 20-ns switching is too slow, 10-pC glitches are unacceptable, and bandwidths in the hundreds of MHz must be controlled, DMOSFET offers the solution.

Fig. 5-20 shows a typical switch connection for ± 10 V analog signal processing. Although the switch conducts equally well in both directions, because feedback capacitance is lower at the drain, it is best that the drain be the output port. This configuration improves both the OFF isolation and the charge injection performance. In this application the gate is driven by a $+20$ V to -10 V square wave.

The equivalent circuit for both ON and OFF states are shown in Fig. 5-21. To remove any possibility of signal path conduction to the body, both the body-source and body-drain *pn* junctions should always be re-

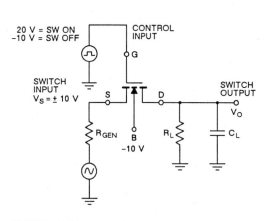

FIGURE 5-20 Normal switch configuration for ± 10 V analog switch. (Courtesy of Siliconix Incorporated.)

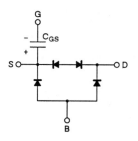

(A) EQUIVALENT "OFF" CIRCUIT

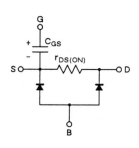

(B) EQUIVALENT "ON" CIRCUIT

FIGURE   5-21 Equivalent circuit for both *(a)* OFF and *(b)* ON. (Courtesy of Siliconix Incorporated.)

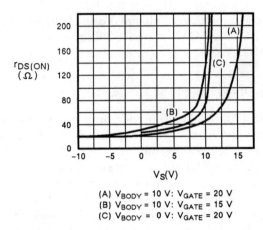

(A) V_{BODY} = 10 V: V_{GATE} = 20 V
(B) V_{BODY} = 10 V: V_{GATE} = 15 V
(C) V_{BODY} = 0 V: V_{GATE} = 20 V

FIGURE 5-22 ON **resistance vs. analog signal voltage. (Courtesy of Siliconix Incorporated.)**

verse biased at a potential greater than the most negative peak signal voltage. For some applications keeping the body at the signal potential is satisfactory.

5-7-1   Switch Characteristics

With a gate-source voltage greater than +5 V, a low-resistance path exists between source and drain. The circuit in Fig. 5-20 exhibits the ON resistance versus analog signal voltage relationships shown in Fig. 5-22.

Numerous applications call for switching a point to ground. For these (current-mode) switches, the source and the substrate are connected to ground, and a TTL-compatible gate voltage of 3 to 4 V is sufficient to ensure switching action.

To achieve minimum distortion, the $r_{DS(on)}$ modulation and the load resistance placed in series with the switch are important factors. For example, if the switch resistance varies between 20 Ω and 30 Ω during a full signal swing and the switch is in series with a 200-Ω load, the total $dR = 4.5\%$, whereas if the load is 100 kΩ, dR is only 0.01%.

Fig. 5-23 shows the effects of V_{SB} and V_{GS} on $r_{DS(on)}$. To maintain a low ON resistance, bias the body to a voltage close to the negative peaks of V_S and use the highest possible gate voltage.

5-7-2   Charge Injection

Charge injection occurs when a gate voltage excursion produces an injection of electrical charge into the analog signal path via the gate-to-drain and the gate-to-source capacitances. Since these devices are not symmetrical, different charges are injected into the source and drain terminals.

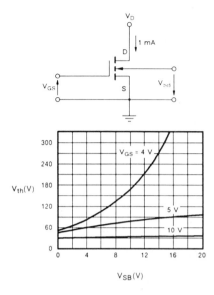

FIGURE 5-23 on resistance vs. body-source voltage. (Courtesy of Siliconix Incorporated.)

Typical parasitic drain-gate capacitances, C_{DG}, and source-gate capacitance, C_{SG}, are on the order of 0.2 pF and 1.5 pF, respectively, for the Siliconix DMOSFETs, type SD210DE.

Switching spikes occur both at switch turn-ON and turn-OFF. At turn-ON, the charge injection effect is minimized by the unusually low signal-source impedance. This low impedance produces a rapid decay of the extra charge in the channel. At turn-OFF, however, the injected charge may be stored in the sampling capacitor, thus creating offsets and errors. The magnitude of these errors is inversely proportional to the magnitude of the holding capacitor.

Fig. 5-24 shows typical charge injection plots. DMOSFET devices provide very low charge injection when compared to other analog switch technologies (e.g., PMOS, CMOS, JFET, and BiFET). However, when the offsets created are unacceptable, they may be minimized using compensation techniques.

5-7-3 OFF Isolation and Crosstalk

For video and VHF switching, the upper frequency limit is determined by the amount of signal that is coupled through the parasitic capacitances. In the OFF state, this signal appears at the switch output when, ideally, no signal should appear.

OFF isolation, when the switch is off, may be calculated as

$$\text{OFF isolation (dB)} = \left| \frac{V_{\text{out}}}{V_{\text{in}}} \right| \tag{5-8}$$

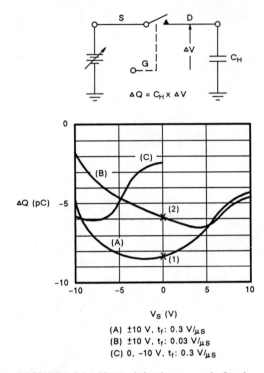

$$\Delta Q = C_H \times \Delta V$$

(A) ± 10 V, t_f: 0.3 V/μs
(B) ± 10 V, t_f: 0.03 V/μs
(C) 0, -10 V, t_f: 0.3 V/μs

FIGURE 5-24   Charge injection scenario for the SD5000 DMOSFET. (Courtesy of Siliconix Incorporated.)

When several analog switches are used to control various high-frequency signals, crosstalk is a key specification. For video applications, the stray signal coupled via parasitic capacitances to the signal of an adjacent channel can cause ghosts and other forms of signal interference. To obtain a high degree of isolation, the circuit needs to be carefully designed to reduce both parasitic capacitive and inductive coupling. Figure 5-25 shows the excellent OFF isolation and crosstalk performance typical of Siliconix DMOSFET devices.

5-7-4   Speed

Because their ON resistance and input capacitance are low, DMOSFET switches are capable of speeds approaching 1 ns. At these speeds the external circuit rather than the FET is responsible for the rise and fall times obtained as was shown in Fig. 2-56, tables 2-1 and 2-2.

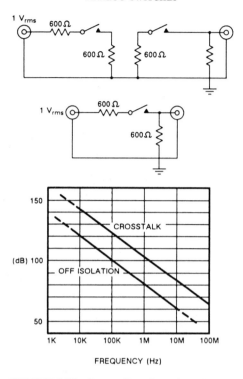

FIGURE 5-25 Crosstalk and OFF isolation of the SD5000 DMOSFET. (Courtesy of Siliconix Incorporated.)

5-8 SWITCHING WITH LOW POWER VERTICAL MOSFETS

The low ON resistance of the vertical MOSFETs when used as an analog switch results in low insertion loss in low-impedance systems, fast charging in sample-and-hold systems, rapid discharge of integrator capacitors, low noise in measuring system, and high accuracy in test systems. Its high current-carrying capability allows transmission of considerable power through the switch, and ease of paralleling without ballast resistors increases this capability. The ability to carry high peak currents is advantageous for driving capacitive lines and quickly charging and discharging capacitors in high-speed A/D converters, S/H circuits, and integrators. Their high OFF isolation (greater than 60 dB isolation at 10 MHz) and less than 500 nA dc leakage provide excellent OFF characteristics. Vertical MOSFETs are fast—a characteristic useful in radar, sonar, and laser applications, where signal bursts must be gated ON and OFF very rapidly. A typical vertical MOSFET can switch 1 A in 4 ns. Their linear ON resis-

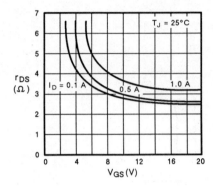

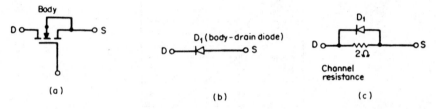

FIGURE 5-26   ON resistance vs. gate-to-source voltage at various drain currents. (Courtesy of Siliconix Incorporated.)

FIGURE 5-27   (a) Schematic symbol of 2N6661; (b) equivalent OFF condition ($V_{GS} = 0$); and (c) equivalent ON condition ($V_{GS} = +10$ V).

tance results in low total harmonic and intermodulation distortion. This is especially important in color video switching, where color purity must be maintained.

Figure 5-26 shows the r_{DS} characteristics of a low-resistance device, the 2N6661. Varying the gate voltage from zero to $+10$ V switches the 2N6661 from an OFF condition to less than 3 Ω ON resistance.

A schematic representation of the 2N6661 and its equivalent OFF and ON circuits is given in Fig. 5-27. It is important to note that the body of the device is internally connected to the source. Diode D_1 is the body-drain p-n junction.

The drain current versus drain-to-source voltage characteristic in the OFF condition (Fig. 5-28) has the appearance of a diode characteristic. The drain current is very low until the drain-source is reverse-biased to about 0.6 V. In the ON condition the channel conductance of about 0.5 mho parallels the diode.

A practical implementation of this device as an analog switch is shown in Fig. 5-29. In the ON condition the gate of the 2N6659 is positive with respect to the source, whereas in the OFF condition the gate voltage is zero. This circuit can take advantage of the 1-A capability and 2-Ω ON resistance of the vertical MOSFET. However, in the OFF state the input

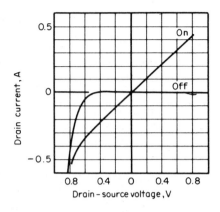

FIGURE 5-28 Small-signal characteristics of 2N6661.

FIGURE 5-29 A simple unidirectional vertical DMOS analog switch.

signal is restricted to positive voltages and should always be greater than the output voltage; OFF isolation is impaired otherwise because of the drain to body diode. Switching times are less than 200 ns, and charge feed-through during the ON to OFF transition is 80 pC with a 50-Ω load. Charge transfer is important in sample-and-hold systems, where an offset voltage of 8 mV into a 0.01-μF load would occur in this case.

To increase the dynamic range to ± 10 V, two 2N6659s are connected in series source-to-source as shown in Fig. 5-30. In the ON condition, both output switches of the DG300 are open, and the gates of both 2N6659s are pulled to $+15$ V. The ON resistance of the switch is now twice the drain-source resistance of a single 2N6659, but the maximum current is still that of a single device. The switch is turned OFF by shorting the gates to the negative supply, reducing V_{GS} to a voltage less than the minimum threshold of 0.8 V. Switch B of the DG300 increases OFF isola-

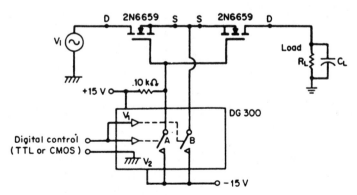

FIGURE 5-30 A general-purpose bidirectional analog switch.

tion 30 dB by shunting the signal leakage path (through the sources of the 2N6659s) to the negative supply.

OFF isolation is shown in Fig. 5-31. The previous problem with the body drain diodes forward-biasing in the OFF condition is now removed, since the two FETs are back-to-back, so one diode is always reverse-biased.

The bidirectional switch has a gate drive which is referenced to a fixed supply. Its ON resistance varies with the input analog voltage because V_{GS} changes (Fig. 5-32). This variation may introduce distortion

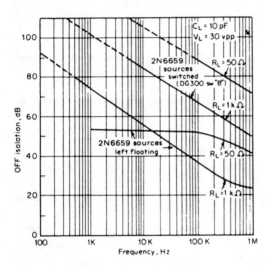

FIGURE 5-31   OFF isolation vs. frequency.

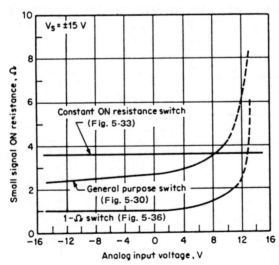

FIGURE 5-32   Small-signal ON resistance vs. analog input voltage.

when driving low-impedance loads such as speakers or transmission lines. To achieve a constant ON resistance, use the circuit shown in Fig. 5-33. In the ON condition, the gates of the 2N6660 are driven by a voltage which tracks the input signal, so gate-to-source voltage is constant and independent of the input signal. No modulation of the ON resistance therefore takes place as the signal level changes. The buffer circuit reduces the total harmonic distortion (THD) from 1.5 percent (no buffer) to less than 0.005 percent (calculated) at 1 kHz, 8 V rms into 50 Ω. The next best analog switch, a 10-Ω DG186, has a 2 percent THD (Fig. 5-34).

The two buffer circuits shown in Fig. 5-35 isolate the input signal

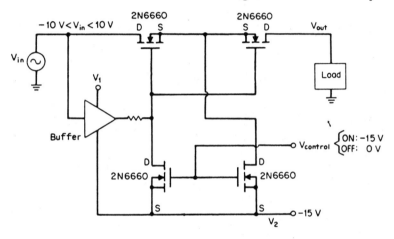

FIGURE 5-33 Low-distortion constant-ON-resistance switch.

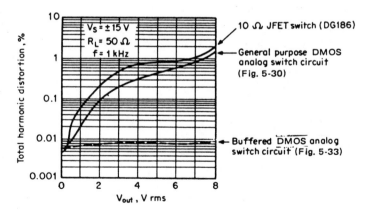

FIGURE 5-34 Distortion improvement using the buffered analog switch.

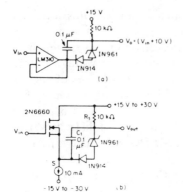

FIGURE 5-35   (a) General-purpose buffer and (b) high-speed buffer.

and use a zener diode to provide a fixed $V_{GS(on)}$ voltage. The general-purpose buffer is flat to 300 kHz and operates on ±15 V or less. The second buffer, a 2N6660 voltage follower, shown in Fig. 5-35b, extends the frequency response to 50 MHz and increases the signal range to ±30 V when operated from ±30-V supplies. The use of a bootstrap in the buffer circuit allows large-signal, low-distortion operation near the positive supply, provided the switch-ON time is small compared to the time constant of R_1C_1.

Lower ON resistance is easily achieved by paralleling devices. For example, three paralleled 2N6659s result in a 1-Ω switch (Fig. 5-36). Paralleling the 2N6659 not only decreases the ON resistance but increases the current-carrying capability to 4.5 A and the linear transfer characteristic to 1.2 A (Fig. 5-37). The voltage range can also be increased. Simply use higher breakdown devices as shown in Fig. 5-38. The 2N6661 has a breakdown of 90 V, allowing up to ±40 V analog capability. However, its ON resistance is also greater (4 Ω versus 1.8 Ω for the 2N6659).

The high-power RF switch shown in Fig. 5-39 has excellent performance up to 50 MHz, with turn-ON and turn-OFF times of less than 50 ns. The isolation at 10 MHz is better than 60 dB with a 20 V p-p input

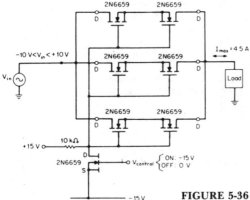

FIGURE 5-36   Ultra-low-resistance switch (1 Ω).

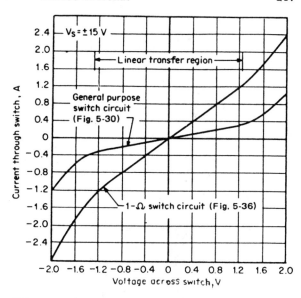

FIGURE 5-37 Large-signal transfer characteristics.

signal; insertion loss is only 1 dB with a 50-Ω load (Fig. 5-40a). The gain versus input power and the two-tone third-order intermodulation product performance curves (Fig. 5-40b) show up to 1 W of power being transferred to the 50-Ω load with a 42-dB intercept point and 1 dB gain compression at 25 dBm input power. The turn-ON time of the switch (Fig. 5-40c) is determined by the passive pullup resistor in combination with the capacitance at the gates of the IRFF113; the negative turn-OFF transient is caused by charge coupling to the output through the output capacitance C_{oss} of the 2N6659.

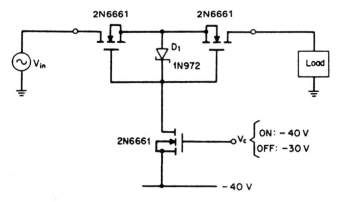

FIGURE 5-38 80-V peak-to-peak analog switch. The zener D_1 prevents the gate-to-source voltage from exceeding $V_{(BR)GSS}$.

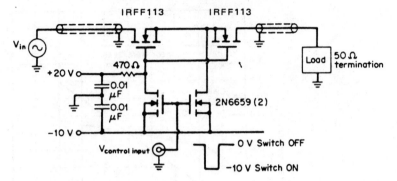

FIGURE 5-39   RF analog switch.

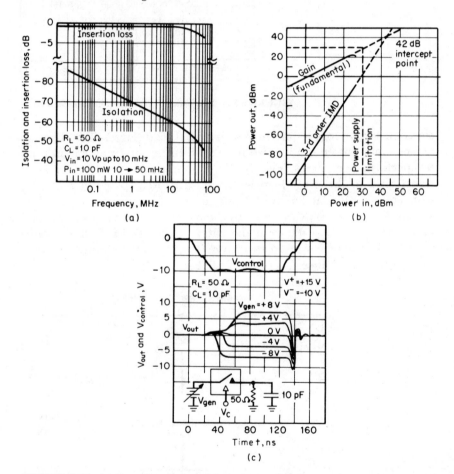

FIGURE 5-40 (*a*) **Insertion loss and isolation vs. frequency of RF analog switch;** (*b*) **gain and two-tone, third-order intermodulation; and** (*c*) **switching response of RF switch into 50-Ω load.**

5-9   DC LEAKAGE CHARACTERISTICS

The ON and OFF dc leakage of the analog switch circuit is a function of the FET switch plus its driver circuit. The calculation of system error due to leakage was presented earlier in dc equivalent circuits. For the JFET switch, the sources of the error-producing leakage currents are shown in Figs. 5-8 and 5-9. In the ON condition, the switch gate is clamped to its source and follows the analog signal. Signal leakage is principally to the -15-V supply due to the OFF leakage of the driver (Q_3 of Fig. 5-4 or Q_2 of Fig. 5-7). It is shown as I_o in Figs. 5-8a and 5-9a. If the signal source resistance R_{SIG} is low, then usually the effect of I_o and $I_{S(off)}$ may be neglected, typical values being less than 10^{-9} A at 25°C. With a value of R_{SIG} of 100 Ω, this magnitude of leakage would result in an error of 100×10^{-9} V.

$I_{D(off)}$ may be a problem in multiplexers where many switches are connected to a common output, such as point A in Figs. 5-1a or b. Then the sum of the switch leakages must be considered. For example, examine the multiplexer circuit of Fig. 5-41a. Assume that one switch is turned ON, $I_{D(off)}$ is the same for all switches, I_o is the leakage of the ON switch driver, and the input conductance of the op amp is zero.

For this case, the error voltage due to leakage current is

$$V_{A(error)} = -[(n-1)\,I_{D(off)}](r_{DS(on)} + R_{SIG}) - I_o R_{SIG} \qquad (5\text{-}9)$$

If, for example, $n = 16$, $r_{DS(on)} = 30\ \Omega$, $R_{SIG} = 100\ \Omega$, $I_{D(off)} = 0.25$ nA, and $I_o = 1$ nA, the error voltage will be

$$V_{A(error)} = -[(15)(0.25)(10^{-9})(30 + 100)] - (10^{-9})(100) \cong -0.59\ \mu\text{V}$$

For high-temperature operation, say at 125°C, $I_{D(off)}$ may increase to

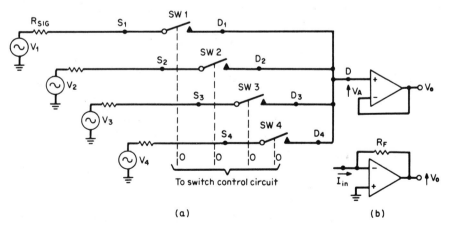

(a)   (b)

FIGURE 5-41   Multiplexer using FET switches.

25 nA, $r_{DS\,(on)}$ may increase to 55 Ω, and I_o to 100 nA. The error voltage then would increase to approximately −68 μV.

Equation (5-9) shows that as the number of switch channels in a multiplexer is increased, the error due to $I_{D(off)}$ becomes more significant. By assuming that $I_o = I_{D(off)}$, $n >> 1$, and $R_{SIG} \cong r_{DS(on)}$, (Eq. 5-9) can be simplified:

$$V_{A\,(error)} \cong nI_{D\,(off)}(R_{SIG} + r_{DS\,(on)}) \qquad (5\text{-}10)$$

If the output of the multiplexer is connected to a current-summing node, such as the inverting amplifier shown in Figs. 5-41 and 5-42, then the error voltage is due to leakage currents into this low-impedance node. Assuming that $R_{SIG} >> r_{DS\,(on)}$, which is typically true for this type of application, then

$$V_{o\,(error)} \cong [\,I_o + (n-1)\,I_{D\,(off)}]\,R_F \qquad (5\text{-}11)$$

when $V_{o\,(error)}$ is the output voltage resulting from switch leakage currents.

5-9-1   Two-Level Multiplexer

If the number of channels n is increased to the extent that the error due to leakage current is excessive, then a two-level multiplexing system may be used. Figure 5-43 shows a two-level system with its equivalent circuit used to determine error due to leakage. For the "voltage-mode" switch, using a high-Z_{in} op amp load, the approximate error voltage due to leakage is

$$V_{A\,(error)} \cong -[(m-1)\,I_{D\,(off)_2}(r_{DS1} + r_{DS2} + R_{SIG})]$$
$$- [\,I_{o2} + (n-1)I_{D\,(off)}](r_{DS1} + R_{SIG}) - I_{o1}R_{SIG} \quad (5\text{-}12)$$

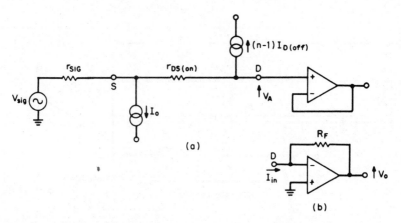

FIGURE 5-42   Equivalent circuit of multiplexer with one switch ON and $n - 1$ switches OFF.

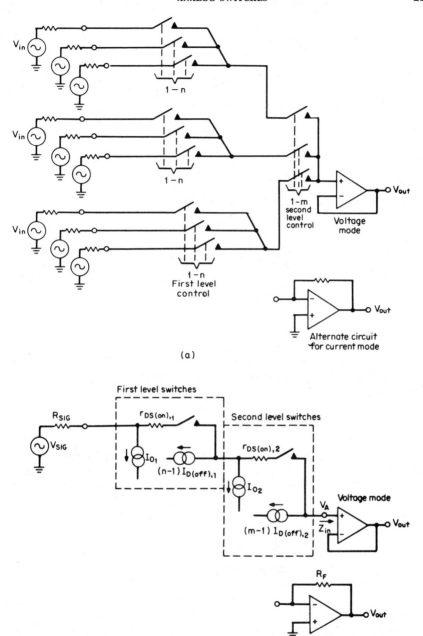

FIGURE 5-43 (a) Two-level multiplexer system and (b) equivalent leakage circuit of two-level multiplexer.

This assumes that $Z_{in} \gg (r_{DS1} + r_{DS2} + R_{SIG})$.

For the current-mode switch, using a low-Z_{in} inverting amplifier, the error due to leakage is

$$V_{o\,(error)} = [\,I_{o1} + (n-1)\,I_{D\,(off)_2} + I_{o2} + (m-1)\,I_{D\,(off)_2}]\,R_F \quad (5\text{-}13)$$

By way of example, compare a 1024-channel single-level system with a 1024-channel two-level system in which $n = m = 32$. Assume that at the maximum operating temperature $I_{D(off)} = I_o = 50$ nA, $r_{DS(on)} = 50\ \Omega$, and $R_{SIG} = 100\ \Omega$. For the single-level system, using Eq. (5-10),

$$V_{A\,(error)} \cong 7.68 \text{ mV}$$

For the two-level system, using Eq. (5-12),

$$V_{A\,(error)} \cong 0.55 \text{ mV}$$

The two-level system has the added advantages of reducing the number of control lines from $n \times m$ lines to $n + m$ lines and reducing capacitance at the output node by approximately the same factor by which leakage current effects were reduced.

5-9-2   MOSFET Leakage

The gate leakage of the MOSFET in a typical analog switching circuit may be neglected when compared to the drain-to-body and source-to-body leakage; thus the driver circuit does not contribute directly to leakage errors. If MOS switches are used in the multiplexer of Fig. 5-41, the equivalent circuit of Fig. 5-44 will be helpful in analyzing the leakage errors.

In the normal operating mode ($V_{SIG} \leqslant V_U$) the leakage currents are due to the reverse-biased source-body and drain-body diodes. The leakage of the ON switch is only slightly greater than $I_{D\,(off)} + I_{S\,(off)}$. Also, for the typical switch, $I_{D(off)} \cong I_{S(off)}$. By inspection of Fig. 5-43 (assuming one switch is ON), the error voltage due to leakage is

$$-V_{A\,(error)} = nI_{D\,(off)}(r_{DS\,(on)} + R_{SIG}) + I_{S\,(off)}R_{SIG} \quad (5\text{-}14)$$

A factor of the MOS switch that may not be obvious is its characteristics in the event of an overvoltage condition where $V_S > V_U$ or $V_D > V_U$. Due to the close proximity of the source to the drain, one can function as an emitter and the other as a collector, with the "body" serving as a base to form a $p\text{-}n\text{-}p$ junction transistor. This means that most of the current injected into the body by a forward-biased source-body junction may be "collected" by a reverse-biased drain. If the source-body or drain-body "diode" becomes forward-biased, the MOSFET no longer provides good OFF source-drain isolation. The analog voltage should not exceed the positive body supply voltage.

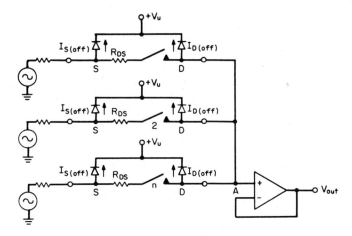

FIGURE 5-44   Equivalent circuit of multiplexer using MOSFET switches.

5-9-3   CMOS Switch Leakage

The CMOS switch shown in Fig. 5-18 has a p-channel MOS and an n-channel MOS in parallel. In its OFF state, the PMOS body is connected to the positive supply and the NMOS body to the negative supply. The results in the PMOS leakage being opposed by the NMOS leakage. The net leakage may be zero at a particular value of drain voltage. The voltage at which this occurs is usually very near the value of the positive supply because in a complementary switch the PMOS is larger than the NMOS and thus has a higher leakage.

5-10   CAPACITANCE AND SWITCHING TRANSIENTS

The capacitance associated with the steady-state ON and OFF states of the FET switch was discussed in the switching high frequency section as it related to signal attenuation and isolation. In this section, we will evaluate the transients which occur during the turn-ON and turn-OFF periods.

By the very nature of its operation the FET requires a certain transfer of charge to change its state from a conducting to a nonconducting mode. The required charge transfer is related to the channel thickness, length, and width and carrier concentration within the channel. Thus, it is related to $r_{DS\,(on)}$ and $V_{GS\,(off)}$. For switching applications, low values of $r_{DS\,(on)}$ and $V_{GS\,(off)}$ are desirable for charge transfer; however, there must be a design compromise between these conflicting parameters.

Increasing channel width will reduce $r_{DS(on)}$ without affecting $V_{GS(off)}$; however, the charge required to turn the device OFF will be increased. Increasing channel thickness will also decrease $r_{DS(on)}$, but increase $V_{GS(off)}$ and thus charge transfer. A reduced channel length results in a decrease in both $r_{DS(on)}$ and charge transfer without an appreciable effect on $V_{GS(off)}$; thus good switching FETs usually have short channels.

5-10-1 Sample-and-Hold Circuit

The charge transfer characteristic is of particular importance in sample-and-hold-type circuits because any charge transferred into or out of the holding capacitor in the transition between the sample-and-hold mode results in an error signal. This characteristic will vary to quite an extent, depending upon various switch and circuit configurations. Few, if any, device data sheets include as a characteristic the charge in picocoulombs required to turn the device ON or OFF. A rough estimate can be had by multiplying $V_{GS(off)}$ and $C_{gs} + C_{gd}$. Caution must be used in comparing units on this basis because the values of C depend upon

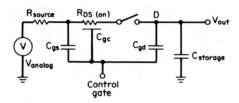

FIGURE 5-45 Equivalent switch circuit for a sample and hold.

bias conditions. Consider Fig. 5-45 as the equivalent circuit for a sample-and-hold circuit. In the switch ON state $C_{storage}$ will charge to the value of V_{analog}. Ideally when the switch is turned OFF, the storage capacitor will hold at whatever value V_{analog} had at the instant the switch was opened. In practice, however, a charge will be transferred from the control gate through the switch capacitance into the storage capacitance. This will result in an offset (error) voltage.

$$\text{Voltage offset} = \frac{\text{charge transfer (picocoulombs)}}{\text{hold capacitance (picofarads)}} = \frac{Q}{C} \quad (5\text{-}15)$$

Using the JFET switch with an integrated driver circuit that was used in the high-frequency example (see Fig. 5-11), we will examine the charge transfer as a function of the analog voltage level. Figure 5-46 shows

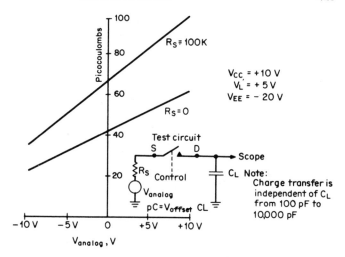

FIGURE 5-46 Typical charge transfer characteristic of the DG181 JFET analog switch.

this characteristic for the analog switch type DG181. In the OFF state the FET gate is clamped to approximately -20 V. In the ON state the gate is essentially at the analog voltage. In going from ON to OFF, then, the control gate has a voltage swing of $V_{EE} - V_{analog}$ volts. The curves of Fig. 5-46 show an increase in charge transfer as V_{analog} is increased. The charge transfer is lower for the case where R_S is zero because a part of the charge is shunted to ground through the signal source. This shows that a lower signal-source resistance results in a lower error. Equation (5-15) indicates that increasing the value of the holding capacitance will reduce the voltage offset resulting from charge transfer.

The JFET construction provides a good compromise between low ON resistance and low coupling capacitance. A comparison of a PMOS switch with the JFET characteristic will show a major difference. The charge transfer characteristic of a PMOS switch type DG172 is given in Fig. 5-47. The switch control circuit is similar to that shown in Fig. 5-18.

As may be seen from the two curves of Figs. 5-46 and 5-47, the charge transfer characteristics for the DG181 and DG172 are similar at $C_L = 100$ pF and $R_S = 0$. It should be noted, however, that the DG172 has a typical ON resistance of 200 Ω—an increase in ON resistance of six times that of the DG181. Values of capacitance greater than 100 pF in the charge transfer characteristics of the DG172 are seen to be inferior to those of the DG181. The major factor which causes the storage capacitance to be value-dependent is the large distributed gate-to-channel capacitance, plus the related circuit time constants.

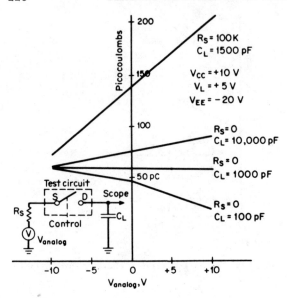

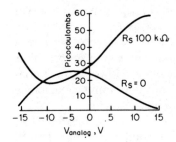

FIGURE 5-47 Typical charge transfer characteristics of the DG172 PMOS analog switch.

FIGURE 5-48 CMOS charge transfer characteristics.

CMOS devices provide an improvement over the JFET and PMOS devices since two gates with complementary control signals are involved. The two resulting charge transfers tend to cancel each other. The transient is therefore greatly reduced but not eliminated. This is shown in Fig. 5-48.

An example of a means of compensating for the charge transfer is shown in Fig. 5-49. In this circuit the charge transfer into the positive and negative inputs of the inverting amplifier is nearly equal; thus the net output voltage resulting from charge transfer is nearly zero. Capacitor C_3 provides a means of adjusting for the nonequality of the two switches in the DG181. (The particular layout of the DG181 results in a slightly higher charge transfer into D_2 than into D_1.)

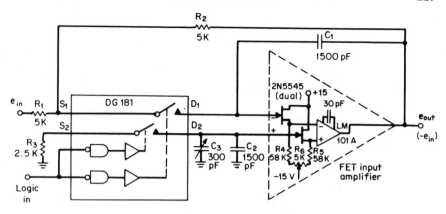

FIGURE 5-49 Improved inverting sample-and-hold circuit.

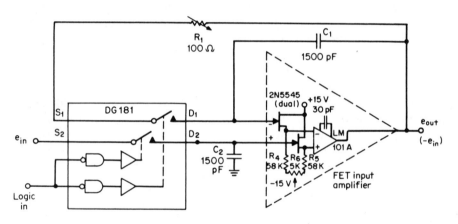

FIGURE 5-50 High-performance noninverting sample-and-hold circuit.

An added feature of this circuit is that the balancing effect of the switch leakage currents results in a reduction in the voltage drop. For designs requiring faster settling times, the noninverting circuit shown in Fig. 5-50 may be used.

5-10-2 Switching Spikes

In switching circuits not involving capacitive loads, the charge transfer during turn-ON and turn-OFF results in a transient voltage or current "spike" or "glitch." The FET switch itself is a high-speed device; the turn-ON and turn-OFF time is essentially that required to accomplish the necessary gate-charge transfer. These times are typically determined by the driver circuit. The faster the switch is toggled, however, the greater will be the peak amplitude of the spike caused by the charge transfer.

The amplitude is also a function of the analog-signal source and load impedances. Typically for the "voltage-mode" switch (Fig. 5-13), R_{SIG} is small so that the transient at the input does not contribute significant error to the system. The apparent load impedance during the switching period will depend upon the status of the other switches in the system. For example, the output spikes will be lower if the switch of Fig. 5-13 is a "make-before-break" than if it is "break-before-make."

The equivalent circuit of two channels is shown in Fig. 5-51. Assuming the switch is an n-channel JFET with driver similar to that used in the DG181 (see Fig. 5-10), the turn-ON output spike will be positive and the turn-OFF output spike will be negative. Switch 1 of Fig. 5-51 has as an output load the parallel combination of R_L, C_L, and switch 2. If switch 2 is OFF, then its loading effect is essentially the capacitance C_{gd2}. Assuming R_{SIG} is low, the spike appearing at the output is due mainly to charge transfer from the driver through C_{gd1}. In turning switch 1 from ON to OFF, part of the charge will flow through $r_{DS(on)}$ and R_{SIG} to ground, so that initially the effective load consists of R_L, C_L, and ($r_{DS} + R_{SIG}$) in parallel.

As the gate becomes more negative, r_{DS} will increase and approach infinity as switch 2 turns OFF; thus ($r_{DS(on)} + R_{SIG}$) will no longer shunt the output. At the same time, C_{gd} is decreasing because V_{DG} is increasing. These two variables plus the lack of a well-defined driver output pulse shape make detailed analysis of the output pulse difficult.

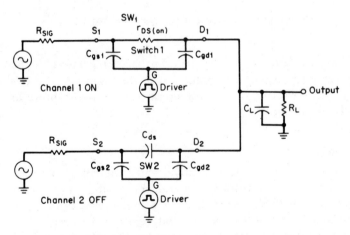

FIGURE 5-51   Equivalent circuit of two-channel multiplexer.

An approximation of the spike output amplitude can be made by utilizing the charge transfer curves shown in Figs. 5-46 and 5-47, the analog voltage level, and the output load. In the simplified equivalent circuit shown in Fig. 5-52, the effective value of C_{gd} is $Q/\Delta V_{GS}$, where Q is the charge transfer in coulombs, ΔV_{GS} is the change in V_{GS} ($V_{analog} - V_{EE}$), and C_{gd} is in farads. For example, let $V_A = 0$. From the

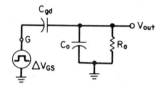

FIGURE 5-52 Simplified circuit for charge transfer analysis.

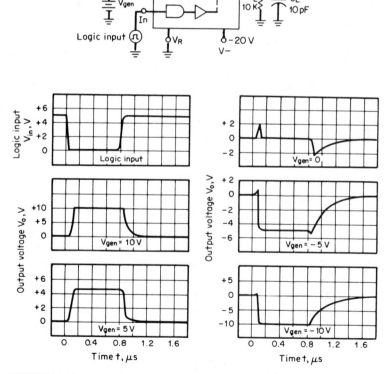

FIGURE 5-53 Typical switching times and transients.

curve of Fig. 5-41, $Q = 40$ pC. Since $V_{EE} = -20$ V, $\Delta V_{GS} = 20$ V and

$$C_{gd\,(off)} = \frac{40 \text{ pC}}{20 \text{ V}} = 2 \text{ pF}$$

If the R_{SIG} is not zero, then the $C_{gd(off)}$ will be increased because the charge lost through the analog source will decrease. From the $R_{SIG} = 100$ kΩ curve of Fig. 5-46, $Q \cong 66$ pC at $V_A = 0$ and

$$C_{gs\,(off)} = \frac{66 \text{ pC}}{20 \text{ V}} = 3.3 \text{ pF}$$

Figure 5-53 shows typical switching transients. Recovery or settling time after the turn-OFF transient is essentially the time constant of $R_o C_o$; however, in a multiplexer application, when one channel is turned OFF, another channel is turned ON, so that the settling time is a function of $r_{DS(on)}$ rather than R_o. Figure 5-54 shows a 32-channel two-level multiplexer with an output settling time of less than 200 ns.

5-11 SIGNAL CONVERSION USING ANALOG SWITCHES

Extensive use of the FET as an analog switch is made in signal conversion systems such as analog-to-digital (A/D) and digital-to-analog (D/A) converters. Increased use of digital techniques to replace analog systems has resulted in the evolution of a multiplicity of converter types. There are still occasions where nonstandard digital codes or analog voltages, a modification of an existing design, or a new design is required. In such cases, FET analog switches as described in this chapter may be useful. Examples are presented below.

5-11-1 Weighted Network Converter

The majority of D-to-A converters available today are based on the weighted network principle. An example of this in binary form is shown in Fig. 5-55. Weighted currents are summed at the output node of the network. The currents are produced from a voltage reference source via an analog switch and a resistor. The value of the resistor depends on the "weighting" of the digit driving it. In the above example, the current produced when the most significant bit is energized is V_{REF}/R. The second most significant bit operation produces a current $V_{REF}/2\,R$, and the third most significant bit operation produces a current $V_{REF}/4\,R$.

In this case the currents are binary weighted. If the resistor values were in decade ratios, the currents would also be in decade ratios. The

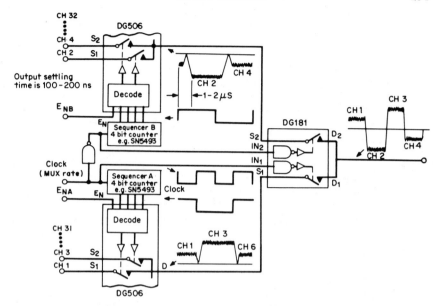

FIGURE 5-54   Practical two-level multiplexer.

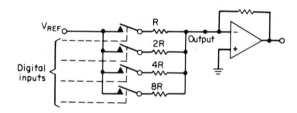

FIGURE 5-55 Binary-weighted resistor D/A converter.

main drawback of this system is the large values of resistance that may be necessary. For instance, for a 10-bit binary-weighted network with $R = 1$ kΩ, the tenth bit has a value 1 k$\Omega \times 2^{10} = 1.024$ MΩ. To maintain the accuracy of the conversion over a temperature range, the resistor ratios must track. This is difficult with resistor ratios of 1000 to 1. Also, the impedance of the network varies with the input code. This can upset the offset and drift of any following amplifier, and hence system accuracy.

5-11-2   Ladder Network Converter

A resistor system that achieves the same "weighting" performance without the large range of values is the binary-weighted ladder network with switches shown in Fig. 5-56. The resistor values have 2:1 ratios.

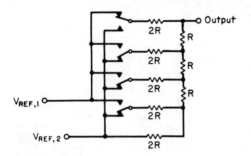

FIGURE 5-56   Ladder network D/A converter.

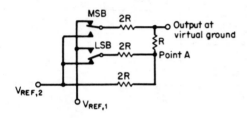

FIGURE 5-57   Operation of two-bit converter based on ladder network principle.

The operation of the ladder network can be understood by considering a simple two-bit converter with V_{REF2} equal to zero volts (shown in Fig. 5-57). With the most significant bit (MSB) equal to logic 1 (V_{REF1}) and the least significant bit (LSB) equal to logic "0" (V_{REF2}), the output current I_z will be $V_{REF1}/2\,R$.

In Fig. 5-58, with MSB = logic 0 and LSB = logic 1 (V_{REF1}), the output current I_z can be calculated using Kirchhoff's equations for the equivalent circuit shown in Fig. 5-59.

$$I_1 = I_2 + I_3 \qquad 2\,RI_2 = RI_3 \qquad I_2 = \frac{I_3}{2} \qquad I_1 = \frac{3\,I_3}{2}$$

$$2\,RI_1 + 2\,RI_2 = V_{REF} \qquad R_3I_3 + RI_3 = V_{REF}$$

hence

$$I_z - I_3 = \frac{V_{REF}}{4\,R} = \tfrac{1}{2}\,\text{MSB}$$

Thus I_2 is now half the value observed in case 1. This approach can be repeated for any number of bits.

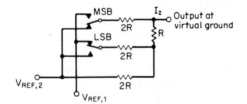

FIGURE 5-58   D/A converter with MSB operating.

Since the ratio of resistor values need only be $2:1$, good temperature tracking can be achieved relatively easily. Furthermore, the ladder network has the advantage of constant output resistance. With reference to Fig. 5-57, the resistance at point A, looking back into the two $2R$ resistors in parallel, is R. From point B, this impedance is in series with R, giving a total of $2R$. Progressing toward the output mode, this $2R$ is parallel with another $2R$ gives R again. Repeating this for any number of bits will show that the impedance of the network at the output is a constant value, R, independent of the input code. This permits the addition of precision gain amplifiers to the converter to produce voltage outputs. The amplifier feedback resistor can be part of the resistor network to produce precise thermal tracking.

5-11-3 Recirculating A/D Converter

Figure 5-60 is but one example of a binary A-to-D converter based upon operational amplifiers and FET analog switches. The principle of operation is relatively simple:

1. The input V_A is sampled when switches S_1 and S_2 are closed and S_3 and S_4 are open.

2. The sample is applied to the amplifier A_2, which is in its reference mode.

   a. If V_A is greater than V_{REF}, the control logic output is a logic 1, and the amplifier will subtract V_{REF} from V_A and multiply the result by 2. This result is then recirculated via switches S_3 and S_2 and treated as a new input.

   b. If the input is less than V_{REF}, the logic is a 0, and the amplifier will be switched to multiply the V_A signal by 2; this output signal is then recirculated as a new input. This would be repeated over a number of cycles up to the resolution capability of the unit, at which time S_1 will open to input a new sample.

The resolution of the converter may be improved by generating more cycles between signal sampling intervals. Switching spikes, which are

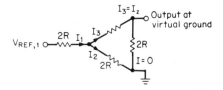

FIGURE 5-59 Equivalent circuit of Fig. 5-58.

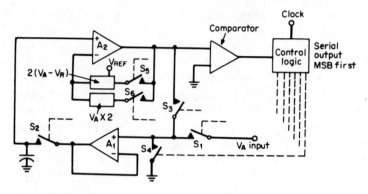

FIGURE 5-60   Recirculating A/D converter.

created when the analog switches in the system are closed, can cause inaccuracies. The overall accuracy is critically dependent on the precision of the analog amplifiers.

The output will be in serial form but can, by the addition of a serial-to-parallel converter, be in parallel form. This design has been fabricated in monolithic PMOS integrated-circuit form. To compensate for MOS amplifier drift, the converter has an auto zero mode which is introduced prior to each input sample. In this mode, the input is grounded via S_4 (Fig. 5-60) and the output added to the analog amplifier reference for the next series of conversion cycles.

5-11-4   Deglitching

When the input digital code changes in a D/A converter, some switches turn ON and others turn OFF. The number of switches operated depends on the exact code change. However, since all types of switches have different turn-ON to turn-OFF responses, there will be rapid changes through many varying codes during the short transitional period. This results in transient spikes or "glitches" (Fig. 5-61). The magnitude of the glitch that can be tolerated depends on the overall system accuracy

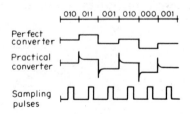

FIGURE   5-61   Sample-and-hold waveforms.

required. A low-pass filter would reduce the glitches, but for high accuracy, a sample-and-hold circuit is often placed after the converter output. The sample-and-hold switch is controlled by a pulse lying within the digit period, such that the sample is taken between transients, when the output is stable.

In circumstances where the input signal to an A/D converter may be changing by a value which is more than the least significant bit, during a complete conversion cycle it is possible to use a sample-and-hold before the converter to provide a stable input.

5-12 CONCLUSION

This chapter has presented some of the characteristics and applications of the FET as an analog switch. A large percentage of all FET applications are for switching. Several reasons for using a solid-state switch in place of a mechanical switch are obvious: high packing density, remote operation and no mechanical linkage, and reliability. These advantages apply to bipolars and FETs. FETs have several additional advantages:

○ High OFF to ON impedance ratios

○ Bilateral operation for large analog signal swing

○ Low power drive due to voltage-control action

Because of its widespread acceptance, the FET has been incorporated, as a switch, into many types of integrated circuits.

6

VOLTAGE-CONTROLLED
RESISTORS AND FET CURRENT
SOURCES

6-1 THE NATURE OF VOLTAGE-CONTROLLED RESISTORS

A voltage-controlled resistor (VCR) is a three-terminal device in which the resistance between two terminals is controlled by a voltage applied to the third terminal. The field-effect transistor is a device whose channel resistance is a function of its gate voltage; thus it can be used as a VCR. In this chapter we will describe the VCR characteristics and applications of n-channel JFETs. The principles can be applied to other types of FETs.

6-2 VCR CHARACTERISTICS OF FETS

Some of the unique resistance properties of the JFET can be deduced from the low-voltage output characteristic curves shown in Fig. 6-1a.

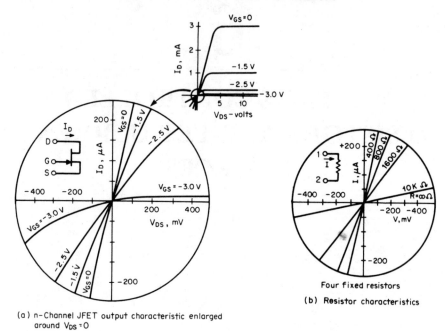

(a) n-Channel JFET output characteristic enlarged
around $V_{DS} = 0$

(b) Resistor characteristics

Four fixed resistors

FIGURE 6-1   Comparison of FET and resistor characteristics.

The slope of the output curve V_{DS}/I_D is a function of V_{GS}; thus the drain-source resistance r_{DS} is controlled by V_{GS}.[1] For comparison the characteristics of four resistors are shown in Fig. 6-1b. It can be seen that r_{DS} of the FET is also a function of V_{DS}; however, this effect is small *if* V_{GS} or V_{DS} is small compared to $V_{GS(off)}$.

The drain-source resistance with $V_{GS} = 0$ and with V_{DS} either at zero or at a very low value is given the symbol $r_{DS(on)}$. A graph of r_{DS} versus V_{GS} normalized to $r_{DS(on)}$ and $V_{GS(off)}$ is shown in Fig. 6-2a. Characteristics of typical devices with various geometry sizes and various values of $V_{GS(off)}$ are shown in Fig. 6-2b. For a given geometry, units with high $V_{GS(off)}$ have the lower $r_{DS(on)}$. Units with the lower $r_{DS(on)}$ will also have the highest I_{DSS}. The relationship between these three parameters is approximated by the equation

$$r_{DS(on)} = -\frac{V_{GS(off)}}{2 I_{DSS}} \tag{6-1}$$

[1]Todd, Carl D., "FETs as Voltage-Variable Resistors," *Electronic Design,* pp. 66–69, September 13, 1965.

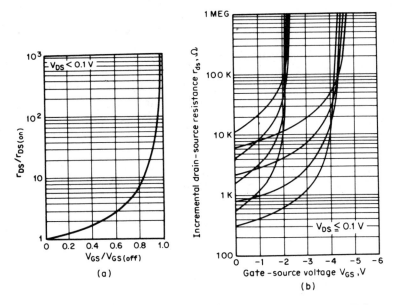

FIGURE 6-2 (*a*) Normalized r_{ds} vs. V_{GS} and (*b*) typical *n*-channel JFET characteristics.

FETs specifically characterized for use as VCRs are offered by several manufacturers. Both *n*-channel and *p*-channel types are available. Figure 6-3 shows that $r_{DS(on)}$ versus $V_{GS(off)}$ relationship of two *p*-channel types and three *n*-channel types offered by Siliconix Inc. These five device types cover a range of $r_{DS(on)}$ values from 20 to 8000 Ω. A FET does not have to be specifically characterized as a VCR to be useful in that function. For example, if a lower value of r_{DS} is needed, the type 2N5432, which has an $r_{DS(on)}$ range of 2 to 5 Ω, may be used.

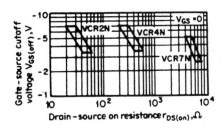

FIGURE 6-3 $r_{DS(on)}$ (drain-source resistance at $V_{DS} = V_{GS} = 0$) varies as an inverse function of $V_{GS(off)}$.

6-3 HOW TO USE THE FET AS A VOLTAGE-CONTROLLED RESISTOR

Figure 6-4 shows a simple voltage-divider attenuator circuit using a FET VCR to provide a means of voltage-controlling the attenuation. As $V_{control}$ is changed from zero to $V_{GS(off)}$ of the FET, r_{DS} changes from $r_{DS(on)}$ to infinity. Letting $g_{ds} = (r_{DS})^{-1}$, the output voltage is

$$V_{out} = V_{in} \frac{R_{load}(1 + g_{ds}R_{load})^{-1}}{R + R_{load}(1 + g_{ds}R_{load})^{-1}} \qquad (6\text{-}2)$$

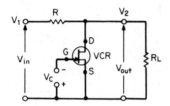

FIGURE 6-4 Simple attenuator circuit.

When V_{GS} approaches $V_{GS(off)}$, g_{ds} approaches zero and the FET causes no signal attenuation. If the load resistance is very large compared to R and $r_{DS(on)}$, Eq. (6-2) can be simplified to

$$V_{out} = V_{in} \frac{1}{1 + Rg_{ds}} \qquad (6\text{-}3)$$

In Chapter 2 [Eq. (2-8)] it was shown that

$$g_{ds} = g_{dso} \frac{V_{GS(off)} - V_{GS}}{V_{GS(off)}} \qquad (6\text{-}4)$$

Using this equation we can express V_{out} as a function of V_{GS}:

$$V_{out} = \frac{V_{in}}{1 + Rg_{dso}[V_{GS(off)} - V_{GS})/V_{GS(off)}]} \qquad (6\text{-}5)$$

Figure 6-5 shows the calculated V_{out}/V_{in} versus V_{GS} for the circuit shown in Fig. 6-4, assuming the constants indicated *and* assuming that the output is small compared to $V_{GS} - V_{GS(off)}$. For a given output voltage, as V_{GS} approaches $V_{GS(off)}$, nonlinearity increases and distortion increases.

6-4 SIGNAL DISTORTION: CAUSES

Figure 6-1a shows that r_{DS} increases as V_{DS} increases in a positive direction. This change in r_{DS} causes the distortion encountered in VCR cir-

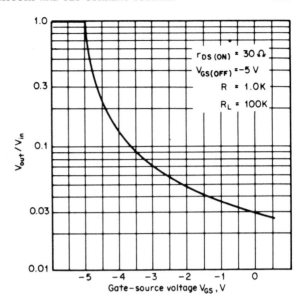

FIGURE 6-5 Calculated V_{out}/X_{in} for circuit of Fig. 6-4.

cuits; it occurs because the channel depletion layer is a function of V_{DG} as well as V_{GS} (see Figs. 1-7, 1-8, and 1-9). The output conductance g_{ds} approaches zero when *either* V_{GS} or V_{GD} approaches $V_{GS\,(off)}$. Also, when V_{DG} swings negative more than a few hundred millivolts, excessive gate current will flow and cause a dc offset voltage at the drain. This is due to the gate-drain *p-n* junction becoming forward-biased.

6-5 REDUCING SIGNAL DISTORTION

Distortion can be reduced and signal-handling capability increased by a simple negative-feedback technique. A portion of the drain signal is coupled to the gate as shown in Fig. 6-6. The desirable effect of this feedback can be explained as follows. If the gate-source voltage is fixed, a positive-going drain voltage tends to decrease the channel thickness,

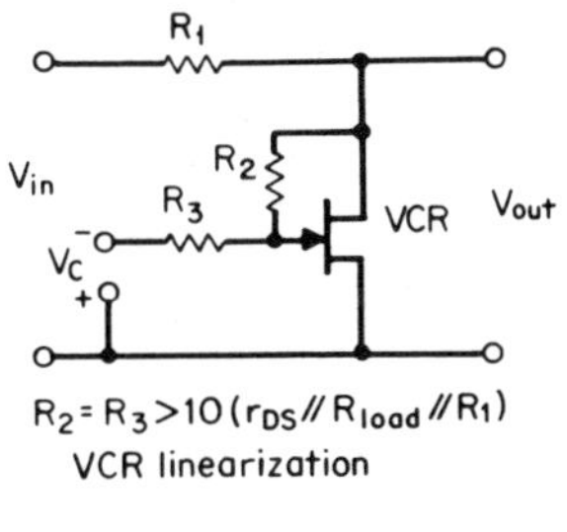

FIGURE 6-6 VCR linearization.

$$R_2 = R_3 > 10\,(r_{DS}\,/\!/\,R_{load}\,/\!/\,R_1)$$

VCR linearization

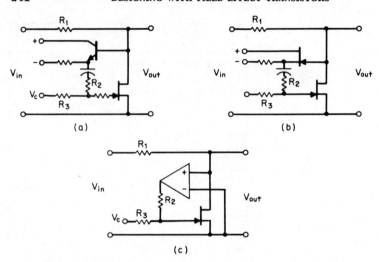

(a)   (b)

(c)

FIGURE 6-7   VCR feedback isolation with voltage followers.

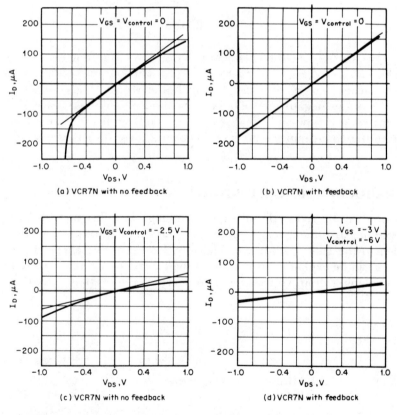

(a) VCR7N with no feedback

(b) VCR7N with feedback

(c) VCR7N with no feedback

(d) VCR7N with feedback

FIGURE 6-8   Output characteristics of VCR attenuators.

thus causing an increase in r_{DS}. By applying a part of the positive-going drain voltage to the gate, the gate-source voltage is made more positive (or less negative) and the thickness of the source end of the channel is increased. This compensates for the positive-going drain-gate voltage. A similiar effect occurs with a negative-going drain voltage.

This method of making V_{GS} a function of V_{DS} decreases the effect that V_{DS} has on g_{ds}, and thus distortion is reduced. The greatest improvement occurs when about $\frac{1}{2} V_{DS}$ is added to V_{GS}; thus in Fig. 6-6 $R_2 = R_3$. To avoid excessive loading of the signal, $R_2 + R_3$ should be much greater than R_1. If the dc control voltage must be blocked from the output, then a capacitor should be placed in series with R_2.

If the coupling of ac control signals from the gate to the output must be avoided, a voltage follower may be added to the feedback loop. Figure 6-7 shows three methods of doing this. Figure 6-8 compares the output characteristics of VCR circuits with and without feedback. It is obvious that linearity is greatly improved with feedback.

It will be noted that a sharp increase in I_D will occur if V_{GD} exceeds about 0.6 V. This drain-to-*gate* current results from the forward-biased drain-gate junction, not from a rapid change in *channel* conduction.

For comparison, a straight line representing a fixed resistor attenuator is superimposed on each of the output characteristics in Fig. 6-8.

6-6   LINEAR GAIN CONTROL WITH VCR

The simple attenuator circuit of Fig. 6-4 provides a V_{out} which has a reciprocal relationship with the control voltage V_{GS}. A linear control of gain can be achieved with the circuit shown in Fig. 6-9. This is a modified ordinary noninverting operational amplifier with feedback. The feedback is controlled by a FET VCR. The gain function of this circuit is

$$A_V = 1 + R_1 g_{ds} \qquad (6\text{-}6)$$

where g_{ds} is the FET conductance ($= 1/r_{DS}$).

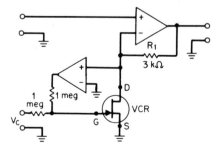

FIGURE 6-9   Amplifier with VCR gain control.

Using Eq. (6-4) we can express the circuit gain as a function of the control voltage V_{GS}:

$$A_V = 1 + R_1 g_{dso} \frac{V_{GS(off)} - V_{GS}}{V_{GS(off)}} \tag{6-7}$$

A minimum gain of unity occurs when V_{GS} is equal to or greater than $V_{GS(off)}$; the FET channel is pinched off and $g_{ds} = 0$. As V_{GS} is made less negative than $V_{GS(off)}$, g_{ds} and thus A_V will increase in a linear fashion. When V_{GS} reaches zero and goes positive, A_V will continue to increase. As has been pointed out before, when V_{GD} exceeds a few tenths of a volt positive, appreciable dc drain current will flow; therefore operating with positive V_{GD} or V_{GS} should be done with caution.

As was shown with the attenuator circuit, distortion in the gain-control circuit is reduced with drain-to-gate feedback. An op amp in the feedback isolates the output from the control V_C. The feedback circuit attenuates the control signal so that $V_{GS} = \frac{1}{2} V_C$. The voltage-gain equation for Fig. 6-7 can be written as

$$A_V = 1 + R_1 g_{dso} \left(1 - \frac{V_C}{2 V_{GS(off)}} \right) \tag{6-8}$$

Figure 6-10 shows the calculated gain function for the circuit of Fig. 6-9. The $r_{DS(on)}$ and the $V_{GS(off)}$ of the FET VCR are assumed to be 30 Ω and -5 V.

If the application requires that the minimum gain be greater than unity, the FET may be shunted with another resistor as shown in Fig. 6-11. The factor R_1/R_2 is added to the gain equation, which becomes

$$A_V = 1 + \frac{R_1}{R_2} + R_1 g_{dso} \left(1 - \frac{V_C}{2 V_{GS(off)}} \right) \tag{6-9}$$

Figure 6-12 shows several types of applications for VCRs. When choosing a FET to function as a VCR, there are several factors which should be considered:

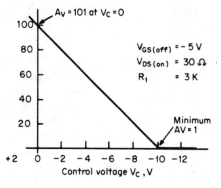

FIGURE 6-10   Amplifier gain vs. control voltage.

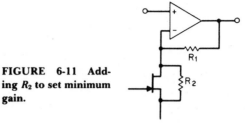

FIGURE 6-11 Adding R_2 to set minimum gain.

1. At low voltages the FET channel conducts equally well in either direction. No *p-n* junctions exist between the source and the drain; thus there is no inherent dc offset voltage.

2. The FET VCR behaves as a linear resistance only for small values of V_{DS}.

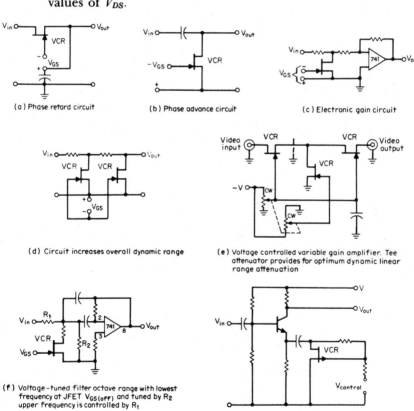

(a) Phase retard circuit

(b) Phase advance circuit

(c) Electronic gain circuit

(d) Circuit increases overall dynamic range

(e) Voltage controlled variable gain amplifier. Tee attenuator provides for optimum dynamic linear range attenuation

(f) Voltage-tuned filter octave range with lowest frequency at JFET $V_{GS(off)}$ and tuned by R_2 upper frequency is controlled by R_1

(g) Wide dynamic range gain control circuit

FIGURE 6-12 FET voltage-controlled resistor applications.

3. The channel conductance g_{ds} is approximately a linear function of V_{GS}.

4. The nonlinearity of r_{DS} increases as V_{GS} approaches $V_{GS\,(off)}$.

5. To improve linearity of the output characteristics and thus lower distortion, apply $\frac{1}{2}$ of V_{DS} to the gate with a feedback circuit.

6. FETs with high $V_{GS\,(off)}$ will have a larger dynamic range than those with low $V_{GS\,(off)}$.

6-7 ANALYSIS OF FEEDBACK LINEARIZED VCRs

The FET as a VCR is typically used in applications where V_{DS} is small and swings $\pm$ about zero. The V-I characteristics near the origin are shown in Fig. 6-1a. In this region, where V_{DG} is less than $-V_p$ and V_{GS} is less than V_p, I_D can be approximated by the quadratic function[2]

$$I_D = \frac{2I_{DSS}V_{DS}(V_{GS} - V_p - \frac{1}{2}V_{DS})}{V_p^2} \tag{6-10}$$

Based on this approximation, the relation between distortion and control range maximum to minimum attenuation will be described. The simple attenuator circuit of Fig. 6-4 less R_L will be used. Most applications can be based on this simple example. The conductance at any point in the first and third quadrants of Fig. 6-1 is

$$G_{DS} = \frac{I_D}{V_{DS}} = -\frac{2I_{DSS}}{V_p}\left(1 - \frac{V_{GS}}{V_p}\right) \tag{6-11}$$

$$-\frac{I_{DSS}}{(V_p)^2}V_{DS} = g_{ds} + \frac{g_{dso}V_{DS}}{2V_P}$$

where g_{ds} is the differential conductance at the origin; when $V_{GS} = 0$, then $g_{ds} = g_{dso}$. The attenuation for the circuit of Figure 6-4 is

$$\frac{V_2}{V_1} = \frac{1}{1 + RG_{DS}} =$$
$$\left[1 + Rg_{ds} + \frac{Rg_{dso}V_1}{2V_p\{1 + Rg_{ds} + Rg_{ds}V_1/[2V_p(1 + Rg_{ds})]\}}\right]^{-1} \tag{6-12}$$

To reduce Eq. (6-12) to a more tractable form, the following inequality is introduced:

$$\frac{V_1 R g_{dso}}{2V_p(1 + Rg_{ds})^2} \ll 1$$

---

[2]von Ow, H. P., "Reducing Distortion in Controlled Attenuators Using FETs," *Proceedings Letters*, IEEE, pp. 1718–1719, October 1968.

so that Eq. (6-12) can now be approximated by the expansion

$$V_2 = \frac{V_1}{1 + g_{ds}R}\left(1 - \frac{Rg_{ds}V_1}{2V_p(1 + Rg_{ds})^2} + \cdots\right) \tag{6-13}$$

Only the second harmonic will be considered for the distortion, since the third is much smaller. For small distortion ($d \ll 1$ and $Rg_{dso} \gg 1$),

$$d = \frac{V_1 Rg_{dso}}{4|V_p|(1 + Rg_{ds})^2} \tag{6-14}$$

If V_2 is held constant,

$$d = \frac{V_2 Rg_{ds}}{4|V_p|(1 + Rg_{ds})} \approx \frac{V_2}{4|V_p - V_{GS}|} \tag{6-15}$$

Figure 6-13 shows a comparison of measured and calculated distortion. If V_{GS} approaches V_p, the above restrictions are violated; the expression for the distortion can no longer be applied. If V_{DG} becomes negative enough for appreciable I_G to flow (the gate-drain junction is forward-biased), the distortion will be higher than predicted. From Eq. (6-15) we get for a prescribed maximum distortion a maximum amplitude as a function of V_{GS}:

$$V_{2\max} = 4d_{\max}|V_p - V_{GS}| \tag{6-16}$$

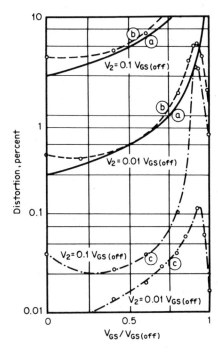

FIGURE 6-13   Distortion as a function of $V_{GS}/V_{GS(off)}$ for two different $V_2/V_{GS(off)}$: (a) theoretical; (b) measured; (c) measured with circuit of Figure 6-14b (© 1968 IEEE; reprinted with permission, from Proceedings of the IEEE, Vol. 56, No. 10, October, 1968, pp. 1718–1719).

For a given d_{max} and V_{2max} the ratio of minimum to maximum attenuation is

$$\frac{A_{min}}{A_{max}} = m = \frac{1 + Rg_{DSS}}{1 + Rg_{DSS}\, V_{2max}/(4\, d_{max}|V_p|)} \approx \frac{4\, d_{max}|V_p|}{V_{2max}} \qquad (6\text{-}17)$$

valid only for $m > 1$. Note that the maximum distortion is reached only for minimum attenuation. Examples:

$$d_{max} = 10 \text{ percent } V_{2max} = 0.001\, V_p m = 400$$

$$d_{max} = 1 \text{ percent } V_{2max} = 0.01\, V_p m = 4$$

Although these relations are only first-order approximations, they give a good estimate of FET attenuator characteristics. The maximum amplitude is proportional to V_p. FETs with high V_p are desirable for attenuator applications. Unfortunately, the majority of commercially available FETs are made with low V_p for use in amplifiers.

There are several means of reducing distortion. By connecting two identical FETs in antiparallel or antiseries, nonlinearities can be cancelled out to a certain extent. A better linearization is possible using one FET with feedback. It has been shown above that the characteristics would be symmetrical if V_{GD} were the control voltage in the third quadrant. If $0.5V_{DS}$ is added to the control voltage, the two voltages V_{GS} and V_{GD} interchange when V_{DS} changes sign:

$$V_{GS} = V_H + 0.5 V_{DS}$$
$$V_{GD} = V_H - 0.5 V_{DS} \qquad (6\text{-}18)$$

Then Eq. (6-18) used in (6-12) gives

$$I_D = \frac{2 I_{DSS}}{V_p^2} V_{DS}(V_H - V_p) \qquad (6\text{-}19)$$

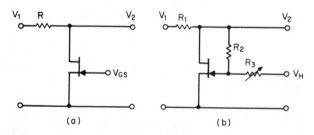

(a)   (b)

FIGURE 6-14 (a) Controlled JFET attenuator and (b) controlled attenuator with "feedback" making characteristics linear and symmetrical.

The resulting characteristic is linear and symmetrical about the origin. The improvement in distortion performance can be seen in Fig. 6-13. A distortion of 12 percent for $V_2 = 0.1V_p$ at $V_{GS} = 0.8V_p$ is reduced through linearization to 0.1 percent. Figure 6-14b shows a possible circuit. The frequency range of the controlled signal must be much higher than that of the controlling signal V_H to keep the direct interference of V_H on V_2 small. R_3 is set for minimum distortion. If V_2 and V_H are in the same frequency range, a high-impedance amplifier must be used. V_2 is at the input; the output is connected to the FET gate. The amplification is approximately 0.5 (adjustable). The control voltage is introduced through a second input so that no direct interference with V_2 occurs.

6-8   FET AS A CONSTANT-CURRENT SOURCE

The ideal "constant-current" source would supply a given current to a load independent of the voltage across the load. In such a case the output conductance of the current source would be zero. Also, ideally the current would be independent of temperature. The drain current of the JFET approaches saturation when the JFET is operated with the gate-drain voltage greater than $V_{GS(off)}$. Under this condition the output conductance g_{ds} is low, and the FET can be used as a current source.

Figure 6-15a illustrates a differential amplifier stage utilizing a FET as a constant-current source to provide a current I_S which is fairly independent of the common-mode voltage. The U401 dual FET is character-

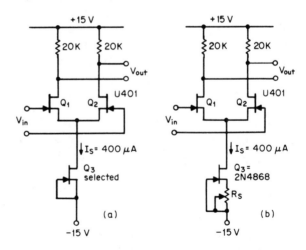

FIGURE 6-15   FET current source for differential amplifier.

ized at $I_D = 200$ μA; therefore the common-source current I_S should be set at 400 μA.

For example, we will assume that the Siliconix Inc. JFET geometry NPA is used for the FET current source Q_3. If a device which has an $I_{DSS} = 400$ μA is selected from this geometry, then its gate can be shorted to its source as shown in Fig. 6-15a. The typical performance curves for the NPA geometry indicate that the selected device would have $V_{GS(off)}$ less than 1 V and an output conductance of less than 2 μmho.

Figure 6-15b includes a resistor R_S in the source of Q_3. This permits a wide range of I_{DSS} values to be used and avoids the necessity of using a selected device. Q_3 could be a standard device such as the 2N4868, which is made with the Siliconix NPA geometry.

The 2N4868 has an I_{DSS} range of 1 to 3 mA and a pinchoff range of -1 to -3 V. The approximate V_{GS} required for I_D is

$$V_G = V_{GS(off)}\left[1 - \left(\frac{I_D}{I_{DSS}}\right)^2\right] \qquad (6\text{-}20)$$

The source-resistor value can be determined by

$$R_S = \frac{V_{GS}}{I_D} \qquad (6\text{-}21)$$

In our example, then, R_S should be adjustable from 840 Ω to 7.4-kΩ.

In the circuit of Fig. 6-15a, if the common-mode input voltage is zero, Q_3 could be replaced with a 37.5-kΩ resistor (neglecting the V_{GS} of the input pair). With the -15-V supply, this would provide the desired 400-μA I_S. If, however, the common-mode voltage varies from, say, $+5$ to -5 V, then I_S would vary from 533 to 267 μA (again neglecting the small V_{GS} change of the input pair). In the case of the FET current source, assuming an output conductance of 1 μmho, the current I_S would change only from 405 to 395 μA.

In the above example with the source resistor, the ±5-V common-mode input signal would cause a 2.66-V peak-to-peak common-mode output. The use of the FET as a current source reduces the common-mode output to 0.1 V peak-to-peak. To achieve a similar low common-mode gain with a resistor, the source supply would have to be increased to -385 V so that the source resistor could be increased to 1 MΩ.

The output conductance g_{ds} of the FET is an important characteristic in its application as a current source. For a given geometry, g_{ds} is a function of I_D and an inverse function of V_{DS}. When adjusted to a given I_D, as in Fig. 6-15b, g_{ds} will be lower for FETs which have lower pinchoff voltage. For example, the performance curves for the Siliconix NPA geometry indicate that units having 5 V, 3 V, and 2 V pinchoff will, respectively, have g_{ds} of 0.75 μmho, 0.58 μmho, and 0.42 μmho, when biased for an $I_D = 2$ mA.

These curves and typical curves for other geometries indicate that the g_{ds}-versus-I_D relationship has the form

$$g_{ds} = g_{dso} \left(\frac{I_D}{I_{DSS}} \right)^n \tag{6-22}$$

where g_{dso} and I_{DSS} are values at $V_{GS} = 0$. This form is similar to the g_{fs}-versus-I_D relationship. The value of the exponent n is a function of the device geometry and of the pinchoff voltage. For the various n-channel geometries evaluated, the exponent n lies between 0.33 and 0.9, with the higher-pinchoff units having higher values for n. However, the higher-V_p devices also have a higher g_{dso}/I_{DSS} ratio.

In the circuit of Fig 6-15b the output conductance of the current source will be lower than simply the g_{ds} of the FET. It is reduced by the degenerative feedback developed across the source resistor R_S. The approximate output conductance including the effect of R_S is

$$g_o = g_{ds} \left(\frac{1}{1 + R_S g_{fs}} \right) \tag{6-23}$$

Since the required value of R_S will be greater for units with higher V_P, the feedback will be greater. This gives some compensation for the higher g_{ds} of the high-V_P units.

In estimating the value of g_{ds} for a given device, it should be remembered that it is a function of both I_D and V_{DS}. The relationships of g_{ds} versus I_D and g_{ds} versus V_{DS} are functions of both the device geometry and $V_{GS(off)}$. It is probably easier to determine g_{ds} at a particular operating point by studying the performance curves given in the manufacturer's data book than to depend entirely upon a formula. Figure 6-16 shows typical performance curves for an n-channel JFET geometry. It can be seen that g_{ds} is a function of I_D, V_{DS}, and $V_{GS(off)}$.

The rate of change of g_{ds} versus V_{DG} increases rapidly as V_{DG} approaches $-V_{GS(off)}$. This indicates another advantage of units with low values of $V_{GS(off)}$; they will perform well at low voltage.

For devices operating at I_{DSS}, such as Q_3 in Fig. 6-15a, the temperature coefficient (θ_I) of I_D will be a function of $V_{GS(off)}$. If the pinchoff voltage is 0.65 V, θ_I will be near zero. Higher-pinchoff units will have a negative θ_I, and lower-pinchoff units will have a positive θ_I. The higher-pinchoff units can be biased to approximately $V_{GS} = V_{GS(off)} + 0.65$ V to achieve zero θ_I. The use of a source resistor as in Fig. 6-15b can improve both the g_o and the θ_I of the current source.

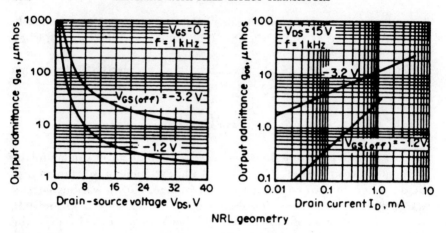

NRL geometry

FIGURE 6-16   FET output conductance characteristics.

6-9   CASCADE FET CURRENT SOURCE

Output conductance can be further reduced by cascading two FETs as shown in Fig. 6-17. In effect this circuit is similar to the current source of Fig. 6-15b, except that as a feedback element R_S is replaced with the output conductance of Q_1. The approximate output conductance

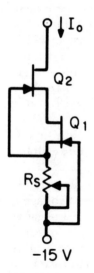

FIGURE 6-17   Cascade for lower g_o.

of the circuit of Fig. 6-17 is given by

$$g_o = \frac{g_{ds2}}{1 + (g_{fs2}/g_{o1})} \qquad (6\text{-}24)$$

where g_{o1} is given by Eq. (6-23). In this circuit g_{ds1} will be higher than might be expected because V_{DS} of Q_1 is only equal to $-V_{GS}$ of Q_2.

6-10 CURRENT-REGULATOR DIODES

The desirable characteristics of the FET as a constant-current source stimulated the development of the current-regulator diode. One version includes an integrated source resistor to improve current control and to lower g_o and θ_l. Figure 6-18 shows the schematic, the symbol, and the equivalent circuit for this type of current-regulator diode. Examples of these units are Siliconix types CR022 through CR530. Three different geometries are used to cover the current range of 220 μA to 4.7 mA.

In Fig. 6-18 the diode I-versus-V characteristic curve is shown, and commonly used symbols and definitions are also given. Note that the drain end of the FET is called the anode and the source the cathode.

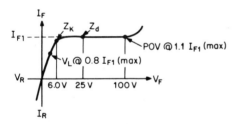

(a) Current-limiter diode V-1 characteristic

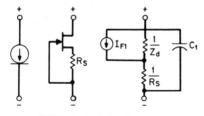

(b) Equivalent circuit

FIGURE 6-18 Current-regulator models and $V\text{-}I$ characteristics (Siliconix CR022-CR530).

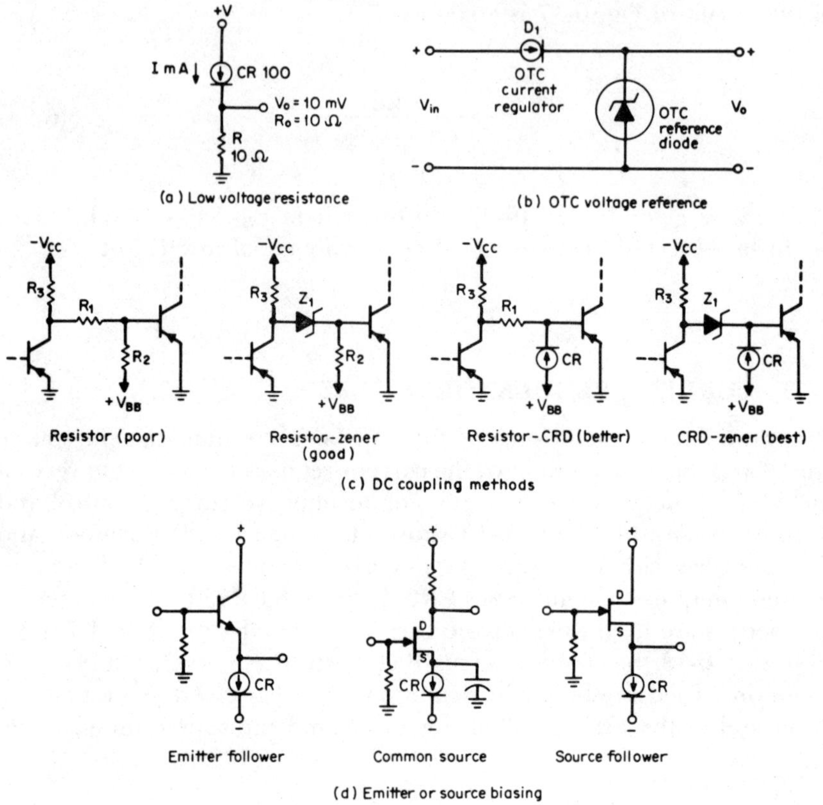

(a) Low voltage resistance

(b) OTC voltage reference

Resistor (poor)

Resistor-zener (good)

Resistor-CRD (better)

CRD-zener (best)

(c) DC coupling methods

Emitter follower

Common source

Source follower

(d) Emitter or source biasing

FIGURE 6-19   Current regulator applications.

The FET current-limiter diode is the electrical dual of the zener diode voltage regulator. The FET diode has low output conductance g_o, while the zener diode has low output resistance r_o.

Figure 6-19 shows several applications of the FET current-regulator diode. Some of the more important applications of current regulators include low-voltage references. Zeners are not available at voltages below 4 V. A precision voltage or millivolt reference is constructed in Fig. 6-19 where the current regulator simply drives a resistor. The output reference voltage is simply determined by $I_F R$.

Another benefit that cannot be obtained with zener references is the low noise of this voltage reference, constructed using the current regulator and a resistor. Neither of these devices is operating in a mode which contributes significant noise, especially at low frequency (which is hard to filter).

Consider the current regulator and resistor as a method of obtaining low-noise regulators even in the normal zener voltage range. One word of caution: the output impedance in this circuit (Fig. 6-19a) is determined by the resistor R.

An excellent solid-state zero temperature coefficient (OTC) voltage reference source is shown in Fig. 6-19b. Here we have combined an OTC zener with an OTC current regulator. Ripple from V_{in} to V_0 in Fig. 6-19b is reduced by more than 120 dB.

A very important application of FET current limiters is in dc coupling between gain stages and level shifting. Figure 6-19c shows methods of dc coupling between two transistor stages.

The use of the zeners and/or current regulators substantially reduces the gain loss otherwise encountered in resistive coupling dc amplifiers.

Differential amplifiers typically utilize current sources in the common emitters or common sources to help achieve the common-mode rejection of these amplifiers. Figure 6-15 shows how the FET current regulator is used in the differential amplifier. For the FET amplifier, common-mode rejection is

$$CMRR_{(V_{CM})} \approx 1 + 2 g_{fs} Z_d \qquad (6\text{-}25)$$

where Z_d = output impedance of current source

g_{fs} = transconductance of the amplifier FETs

The output impedance of the current source is the key to improving common-mode rejection.

Single-ended transistor amplifier performance can be improved using current regulators. Figure 6-19d shows three different amplifiers using emitter and source current-regulator biasing. In the emitter follower the current regulator significantly increases the input impedance to the circuit and develops a gain closer to unity; less obvious is a lower transistor dissipation when supplying a heavy external load.

In the FET common-source amplifier the dc bias point is well defined using the FET current regulator. This allows zero temperature coefficient biasing of the amplifier. The same bias temperature stability is achieved in the source-follower circuit.

Another important application of the current regulator is in high-speed bootstrapping circuits. Figure 6-20 is a high-speed line driver using power MOSFETs. In this circuit the current regulator CR047 minimizes power dissipation, at the same time maximizing speed. When Q_1 turns OFF, the CR047 discharges the parasitic capacitance rapidly in a controlled manner, turning ON Q_2 by enhancing the available gate voltage. Once the voltage across the current regulator drops below the limiting voltage V_L, it becomes very low resistance; thus the gate is tied to the top of the bootstrap capacitor.

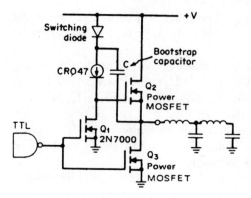

FIGURE 6-20 Application of current regulator (CR047) in a high-speed line driver.

6-11 WAVEFORM GENERATION USING CURRENT REGULATORS

A simple linear sawtooth generator can be constructed as shown in Fig. 6-21a. The use of the current regulator in series-opposing fashion makes possible the generation of high-quality triangular waves from a sine- or square-wave source, as Fig. 6-21b illustrates. Square-wave drive results in a better waveform at the zero crossings. Output frequency is identical to the input frequency.

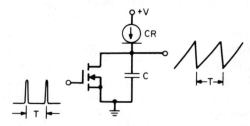

(a) Linear sawtooth generator

(b) Triangular waveform generator or integrator

$$V_{opp} = \frac{I \times t}{C}$$

FIGURE 6-21 Waveform generation.

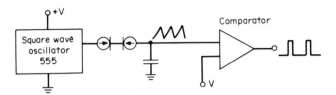

FIGURE 6-22 Pulse-width modulator.

A very useful application of the triangular-waveform generator is in pulse-width modulators, where the triangle wave is compared against a feedback voltage. The output of the comparator is an accurate pulse-width-modulated representation of the analog feedback voltage (Fig. 6-22).

A conventional clipper circuit or square generator is shown in Fig. 6-23a. One disadvantage is the poor dynamic resistance encountered in the low-voltage zener, which results in the fairly poor output waveform shown. The simple addition of series-opposing current regulators, shown in Fig. 6-23b, results in a flat-top output waveform and an efficient circuit. The output waveform is $\pm(V_z + 0.7)$ V.

As shown in Fig. 6-24, FET current-regulator diodes can be connected in series to extend the voltage range, in parallel to extend the current range, and in series-opposing to function as a bidirectional current limiter.

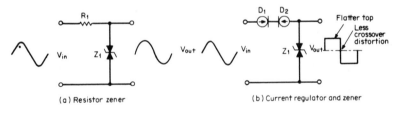

(a) Resistor zener

(b) Current regulator and zener

FIGURE 6-23 Square-wave generator or clipper.

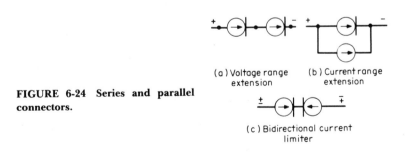

(a) Voltage range extension

(b) Current range extension

(c) Bidirectional current limiter

FIGURE 6-24 Series and parallel connectors.

6-12   POINTS TO REMEMBER WHEN CHOOSING FETs AS CURRENT SOURCES

1. Shorting the gate to the source of a FET makes a simple current limiter.

2. Put a resistor in series with the source of a FET (Fig. 6-18) and you have a low-temperature-coefficient, high-output-impedance current source.

3. Place two FETs in a totem pole configuration (Fig. 6-17) and you increase the output impedance of the current source by more than an order of magnitude.

4. Applications requiring two-terminal current regulators without an internal supply are made very simply using depletion-mode FETs.

BIBLIOGRAPHY

Gosling, W.: "Voltage Controlled Attenuators Using Field-Effect Transistors," *IEEE Trans Audio,* **AU-13,** pp. 112–120, September–October 1965.

Sevin, L. J.: *Field-Effect Transistors,* McGraw-Hill, New York, 1965.

Sherwin, J. S.: "Voltage Controlled Resistors (FET)," *Solid State Design,* pp. 12–14, August 1965.

7

POWER FETs

7-1 INTRODUCTION

Field-effect transistors, like their bipolar cousins, were for the first several years of their existence useful only at low (<1 W) power levels. While they possessed many theoretical advantages over their bipolar counterparts, the practical limitations in manufacturing high-power devices precluded FETs from competing with bipolar transistors and SCRs in power applications. The major limitation was that FETs were strictly horizontal devices; that is, their channels were parallel to the chip surface, so that their current densities were much lower than the bipolars' (which utilized vertical current flow). For a given current, the FET chip area had to be considerably larger, which meant a lower yield and higher cost. Medium-power FETs were therefore more costly to fabricate than their bipolar counterparts, while high-power FETs were nearly impossible.

Several new technologies have been developed to increase current density and allow production of high-voltage, high-current FETs. Two of these technologies, vertical DMOSFET and the static induction transistor (SIT), are presently in production; the vertical DMOS is more popular and widespread. Quickly coming on scene is the insulated-gate bipolar transistor (IGBT) that may offer even greater current and voltage capabilities than the now-popular vertical DMOSFET.

Vertical power FETs have unique characteristics and capabilities that are not offered by bipolar power transistors. For example, like the FETs described in this book, they are controlled by a gate-source voltage rather than by base current which means that drive circuits can be simplified. Switching times are also much faster than bipolar devices, particularly as there is no equivalent to saturation and there is no storage time (*turn*-OFF time, Yes; *storage* time, No)! Recent advances in technology have also led to the introduction of very low ON-state resistance devices exhibiting saturation voltages lower than any other semiconductor.

7.2 THE POWER MOSFET STRUCTURE

The structure of a vertical power MOSFET differs from any other type of MOSFET in that the current flows vertically through the silicon die instead of laterally. Fig. 7-1 shows a simplified cross-section through the basic cell. Early in the history of power FETs the V-groove VMOS, shown in Fig. 7-2, was the first, somewhat successful, commercial device. Since 1980, however, the vertical DMOS structure, shown in Fig. 7-1, has predominated.

A double-diffused structure consists of a sequentially introduced set of impurities into the epitaxy that for an n-channel MOSFET consists of first a deep p-doped region followed by a more heavily doped n-region. The channel length is controlled by careful monitoring and control of the doping levels and the subsequent diffusion cycle which must consider both time and temperature. One obvious difference between the DMOS structure and the conventional planar MOS structure is the channel length determined by the difference between the two sequential diffusions. Unlike the conventional planar MOS where the channel length is achieved through the photolithographic process and limited to 4 to 5 μm,

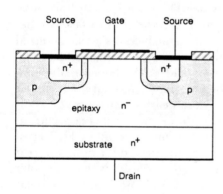

FIGURE 7-1 Idealized cross-sectional view of a vertical power MOSFET.

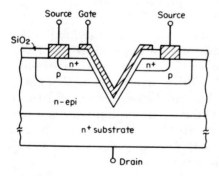

FIGURE 7-2 Cross section of a V-groove MOSFET channel.

the DMOS channel length can be reproducibly controlled to values in the 1 to 2 μm range.

A less obvious difference, but one of paramount importance, is what happens to the *pn* depletion field that formerly forced the construction of a long channel in the conventional planar MOS. In the DMOS process, the body region (for the *n*-channel MOSFET, the *p*-doped area) is more heavily doped than the *n*-drain (epitaxy) region. Consequently, the depletion field extends further into the drain region than into the body region when a reverse bias is placed across the drain-to-body (*np*) junction. This not only allows significantly higher voltages to be placed across the junction without forcing a longer channel, it also maintains a fixed threshold voltage with varying drain-to-source potential. These two differences allow a MOSFET that has both a short channel length and the ability to withstand a higher drain potential without fear of punchthrough.

The amount of voltage that a vertical power MOSFET may withstand is a function of both the resistivity and thickness of the epitaxy region. Likewise, as this epitaxy resistivity increases, thus allowing higher withstanding voltages, the overall ON resistance of the MOSFET rises, but not proportionally, as Eq. (7-1) shows.

$$r_{DS(\text{on})} = K \, V_{(BR)DSS}^{2.5} \qquad (7\text{-}1)$$

Equation (7-1) identifies the fundamental problem with the vertical power DMOSFET: as the breakdown voltage rises, the ON resistance rises more rapidly, thus severely limiting the current capacity of the high-voltage devices. Because the IGBT does not rely upon a high resistivity epitaxy to support its operating voltages, it may supplant the vertical power DMOSFET in high power applications.

For a vertical power DMOSFET to handle high levels of current while withstanding high breakdown voltages, K is a proportionally constant based on, among other things, the chip size. Multiple paralleling of DMOSFET elements on a single chip obviously reduces the total resistance and with it, because of the increased chip area, improves the thermal resistance and allows for greater current handling.

For any power FET, device operation is initiated by the application of a bias voltage to the polysilicon gate. This induces an inversion layer at the surface of the body region (p) between the source (n^+) and the drain (n^-) regions. The depth of this inversion layer (known as the channel) increases with increasing gate bias up to a certain limit, controlling current flow between the drain and the source. This current flow is also controlled by the applied drain-to-source voltage where this voltage is relatively low. As the drain-to-source voltage increases, the current saturates and becomes dependent only on the gate-to-source voltage.

7-3 THE VERTICAL JFET TECHNOLOGY

Figure 7-3 is the channel cross section of the vertical JFET, another type of power FET which, like DMOS, has a high power density and high breakdown voltage capability. In the vertical JFET (V-JFET), current flows vertically from the drain through the channel between the gate fingers and into the source. A negative gate voltage $-V_{GS}$ causes the depletion region to reach further into the channel, reducing its width and constricting current flow. Ultimately, when the gate is sufficiently negative, the depletion regions from adjacent gate fingers touch, and current flow is stopped altogether. However, because the channel is short, it is not pinched off by increasing the drain voltage, as it would have been in a conventional longer-channel JFET.

The output characteristics of one type of JFET (2SK60) are shown in Fig. 7-4. The drain current does not saturate with increasing drain

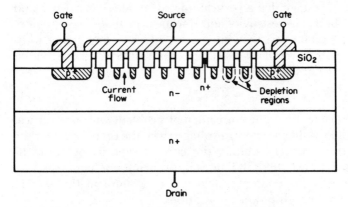

FIGURE 7-3   Cross section of a vertical power JFET.

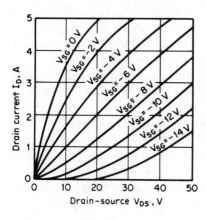

FIGURE 7-4   Output characteristics of a 2SK60 vertical power JFET.

voltage but continues to be a function of drain voltage. The characteristics are similar to those of a triode vacuum tube; i.e., $\Delta I_D/\Delta V_{DS}$ is fairly large, and the voltage amplification factor g_{fs}/g_{ds} is quite low—on the order of 5.

The vertical JFET is necessarily a depletion-mode device (ON when $V_{GS} = 0$), while DMOS are typically enhancement-mode devices (OFF for $V_{GS} = 0$), although they could be made as depletion-mode devices. A second difference is that the p^+ gate diffusion of the V-JFET has a higher resistance than does the polysilicon gate used for DMOS, so the vertical JFET does not achieve the high-frequency response and fast switching times of which DMOS are capable. On the other hand, V-JFETs have a somewhat higher linearity than do other types of FETs, and are therefore suited for low-frequency amplifiers. To date, V-JFETs have primarily been used in high-quality audio power amplifiers.

7-4 POWER DMOS CHARACTERISTICS

The output characteristics of the DMOS type 2N6661, plotted in Fig. 7-5, are similar to those of a conventional MOSFET with these exceptions: The vertical scale is amperes rather than milliamperes, the output conductance g_{os} is low (the curves are flat rather than sloping) because of the buffering effect of the epi region, and the g_{fs} is constant (the lines are evenly spaced) above 0.25 A. The constant g_{fs}, a characteristic of short-channel devices, is due to velocity saturation of the electrons in the channel. Above a certain threshold, increasing the electric field intensity does not increase the drift velocity. The g_{fs} of a conventional (long-channel) MOSFET, on the other hand, is proportional to the gate voltage; drain current is therefore proportional to V_{GS}.

Of the advantages that DMOS has compared to bipolars, many are well

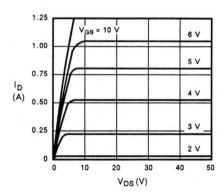

FIGURE 7-5 Output characteristics of the 2N6661. (Courtesy of Siliconix Incorporated.)

known in small-signal applications. Some that are apparent at high power levels are:

1. High input impedance and low drive current (typically less than 100 nA). The dc beta of a DMOS device (the output current divided by the input current) is therefore over 10^9. Since the resultant drive power is negligible, DMOS will directly interface to medium-high-impedance drivers such as CMOS logic or optoisolators.

2. No minority-carrier storage time. DMOS is a majority-carrier device—its charge carriers are controlled by electric fields, rather than the physical injection and extraction (or recombination) of minority carriers in the active region. The switching delay time is small, several nanoseconds, and is caused primarily by external parasitic elements (series gate inductance). The 2N6661, for example, is capable of switching 1 A ON or OFF in 4 ns, about 10 to 200 times faster than a bipolar.

3. No secondary breakdown or current hogging. Since the temperature coefficient of the DMOS drain current is negative (a bipolar's is positive), DMOS draws less current as the device heats up. If the current density tends to increase at one particular point of the channel, therefore, the temperature rises and the current decreases. The current automatically equalizes throughout the chip, so no hot spots or current crowding—which eventually leads to secondary breakdown in a bipolar—can develop. Similarly, current is automatically shared between paralleled devices so no ballasting resistors are needed.

7-4-1 Gate Threshold Voltage

The gate threshold voltage, $V_{GS(th)}$, provides a measure of the voltage required to initiate turn-ON. It is the gate bias that is required to provide a specified drain current that is above the leakage current level, but very low when compared to the normal operating current level of the device.

Gate threshold is generally specified as a range with both minimum and maximum limits. For any value of V_{GS} below the minimum limit, the device will be OFF; for a V_{GS} above the maximum limit, the device will be ON so that the current flow will be at least that at which $V_{GS(th)}$ is specified. The threshold voltage exhibits a negative temperature coefficient, decreasing by approximately 5 mV/°C as Fig. 7-6 identifies for the 2N6661.

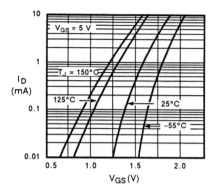

FIGURE 7-6 Temperature effects on threshold voltage for the 2N6661. (Courtesy of Siliconix Incorporated.)

7-4-2 ON-State Resistance

With a gate-source voltage of approximately 10 V, power MOSFETs behave as resistors. Devices rated at high currents have lower ON-resistance than low-current devices. For example, the 2N7000, rated at 200 mA, has an $r_{DS(on)}$ of 5 Ω, whereas the 2N7001, packaged in the SOT-23 surface mount, rated at 45 mA has an $r_{DS(on)}$ of 45 Ω.

The low ON resistance is achieved through paralleling many cells on a single chip. Modern design techniques use a silicon-gate matrix and a surface metalization which connects the source sites of all the cells. The underneath side of the chip is the common drain. Figure 7-7 shows how these cells may be interconnected to achieve the desired power handling.

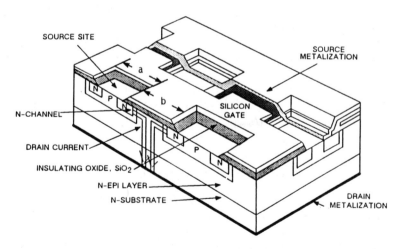

FIGURE 7-7 Idealized cut-away view of a vertical power MOSFET showing the cell positions and major current path. (Courtesy of Siliconix Incorporated.)

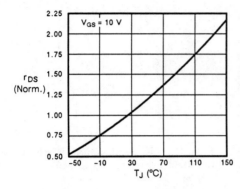

FIGURE 7-8   ON resistance vs. junction temperature. (Courtesy of Siliconix Incorporated.)

As the power MOSFET heats up, $r_{DS(on)}$ increases. There is no simple relationship between $r_{DS(on)}$ and T_j since the resistance consists of a number of components each with different temperature coefficients. However, we see that the temperature coefficient of $r_{DS(on)}$ progressively increases as the breakdown voltage increases because of the increasing role played by the epitaxy resistance. Figure 7-8 illustrates the typical relationship between ON resistance and junction temperature for the 2N6661.

7-4-3   ON-State Parameters

When the gate bias is significantly in excess of threshold the transistor will turn-ON full and ON-state drain current, $I_{D(on)}$, will flow. Under measurement conditions for $I_{D(on)}$, the drain-source voltage, V_{DS}, will be defined. The product of V_{DS} and $I_{D(on)}$ (the power dissipated) will cause the chip temperature, T_j, to rise; thus $I_{D(on)}$ is generally specified under pulsed conditions to ensure that T_j remains close to T_A.

A major contributor to the power dissipated in a power MOSFET is the drain-source ON-state resistance, $r_{DS(on)}$. This is measured under conditions of high gate bias and a defined drain current which is usually about half the drain current ($I_{D(on)}$) rating.

7-5   GENERAL SWITCHING APPLICATIONS

The high input impedance and high speed of DMOS are desirable switch characteristics. DMOS will interface any driver capable of a 5- to 15-V swing to nearly any load requiring several amperes of current.

The basic switching performance of the 2N6661 is shown in Fig. 7-9, while the corresponding test circuit is shown in Fig. 7-10; 40 pF is connected in series with the 2N6661 gate to better match it to the 50-Ω

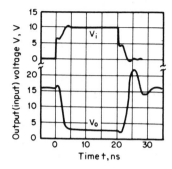

FIGURE 7-9 Switching performance of the 2N6661.

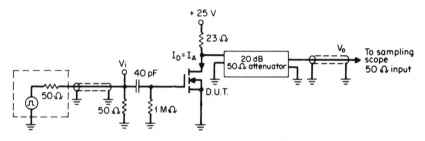

FIGURE 7-10 Switching test circuit for the 2N6661.

source. The 2-ns turn-ON and turn-OFF delay is caused by the input capacitance charging and discharging through the equivalent series inductance of the package and test jig.

CMOS logic, such as the 34011 gate, may be used as a driver for the 2N6661 in Fig. 7-11. A logic LOW to the input of the 34011 turns the 2N6661 ON ($V_{GS} = 10$ V), while a logic HIGH turns the device OFF ($V_{GS} = 0$). The steady-state power dissipated by the circuit, exclusive of load current, is a maximum of 55 μW (0.15 μW typ).

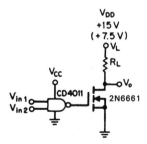

FIGURE 7-11 A CMOS gate driving the 2N6661.

Figures 7-12 and 7-13 depict the dynamic performance of the circuit shown in Fig. 7-11 when the load is 25 Ω. $V_{DD} = 15$ V when the logic supply voltage is 10 or 15 V and 7.5 V for $V_{CC} = 5$ V. The turn-ON and turn-OFF times when $V_{CC} = 10$ V are about 60 ns. Increasing V_{CC} to 15 V decreases t_{on} and t_{off} to 50 ns. Decreasing V_{CC} to 5 V increases the switching times to 120 ns. The input and Miller capacitances of the 2N6661 present a load of typically 65 pF to the 34011 driver.

The switching time is decreased when several CMOS gates are paralleled to increase drive current to the 2N6661. For example, when four 4011 gates are paralleled and V_{CC} is 15 V, switching times are about 25 ns—most of it propagation delay through the 4011 (Fig. 7-14).

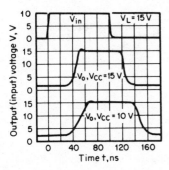

FIGURE 7-12 Switching performance of the 2N6661 driven by the CMOS gate of Fig. 7-11.

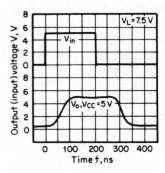

FIGURE 7-13 Switching performance of the 2N6661 with a 5-V CMOS logic drive.

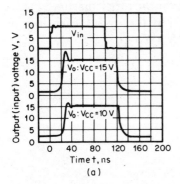

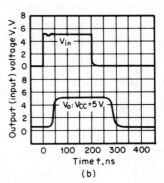

FIGURE 7-14 Switching performance of the 2N6661 driven by four parallel 4011 gates.

To further decrease switching times, additional peak drive current is needed to charge and discharge the gate input capacitance of the 2N6661. One solution is to use a MOS clock driver such as the MH0026. It is designed to deliver high peak currents into capacitive loads and to translate TTL levels into 15-V swings. Figure 7-15 is a typical switching circuit whose characteristics are shown in Fig. 7-16.

DMOS will also interface to standard TTL, but a pullup resistor is needed to ensure sufficient gate enhancement (see Fig. 7-17). If no pull-up resistor is used, the enhancement of the DMOS will be about 3 V, and the DMOS saturation current will be only about 200 mA. On the other hand, with 5 V provided to the gate by the pullup resistor, the 2N6661 saturation current will exceed 500 mA—which is adequate for many applications.

If a higher saturation current or a lower ON resistance is needed, the gate drive voltage must be increased. Figure 7-18 shows how to

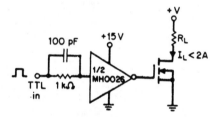

FIGURE 7-15 Driving the 2N6661 with a MOS clock driver.

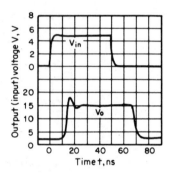

FIGURE 7-16 Switching performance of the 2N6661 driven by a MH0026 MOS driver.

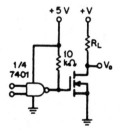

FIGURE 7-17 Driving the 2N6661 with standard TTL.

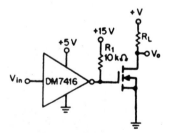

FIGURE 7-18 Open-collector TTL is used to provide greater enhancement to the 2N6661.

use open-collector TTL with a 10- to 15-V pullup. Turn-ON time will be a function of the value of R_1, since it provides current to charge the input capacitance of the 2N6661. If a faster turn-ON time is needed, R_1 must be reduced, which may result in excessive power dissipation when the 2N6661 is OFF. To solve this problem, use the totem pole drive circuit shown in Fig. 7-19—it drives the 2N6661 with an emitter follower with performance as shown in Fig. 7-20.

A second method of interfacing TTL to DMOS, a bipolar level shifter, is shown in Fig. 7-21. The 2N5130 amplifies the TTL output pulse and provides up to 15 V of enhancement to the 2N6661. The AM686 is a high-speed comparator, although other comparators—or for that matter, a TTL gate—could be used. For a faster turn-ON time than that shown in Fig. 7-22, or for a lower power dissipation in the OFF state, use the totem-pole driver.

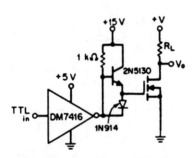

FIGURE 7-19 A "totem pole" drive increases switching speed and reduces dissipation.

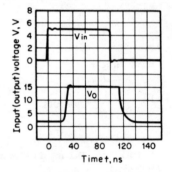

FIGURE 7-20 Switching performance of the 2N6661 driven by the "totem pole" open-collector TTL driver (Fig. 7-19).

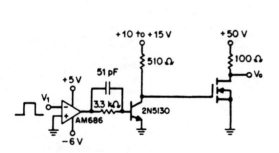

FIGURE 7-21 High-current interface.

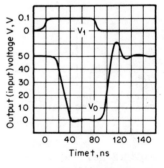

FIGURE 7-22 Performance of the high-current interface.

Interfacing DMOS to ECL is not quite as straightforward, since ECL levels are inherently incompatible with DMOS drive requirements, but level shifting is still relatively easy. In Fig. 7-23, the 2N6661 is used to increase the voltage and current capability of an ECL-compatible peripheral driver, the 75441. An alternative circuit, Fig. 7-24, uses discrete components to translate ECL levels into the 0- to 10-V swing required for DMOS. Switching times of this circuit are less than 40 ns into 50 Ω.

When driving capacitive loads, such as cables or data buses, an active pull-up is required to deliver current into the load. The high-speed line driver shown in Fig. 7-25 uses a second 2N6661, with an inverter, to

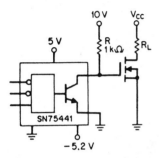

FIGURE 7-23 Using DMOS to buffer the output of an ECL-compatible peripheral driver.

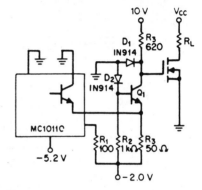

FIGURE 7-24 A discrete ECL-to-DMOS interface circuit.

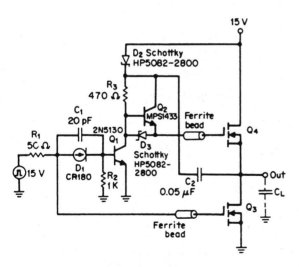

FIGURE 7-25 High-speed line driver.

provide up to 3 A and switch 15 V across 1000 pF in less than 30 ns (Figure 7-26).

Ease of drive, ruggedness, lack of secondary breakdown, and fast switching speeds make DMOS well suited for switching power to a variety of loads, some of which are shown in Fig. 7-27. The fast switching of

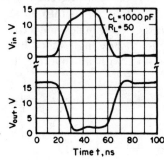

FIGURE 7-26 Performance of the high-speed line driver.

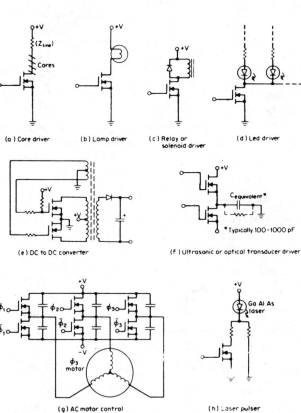

(a) Core driver

(b) Lamp driver

(c) Relay or solenoid driver

(d) Led driver

(e) DC to DC converter

(f) Ultrasonic or optical transducer driver

(g) AC motor control

(h) Laser pulser

FIGURE 7-27 Several typical DMOS applications.

DMOS is especially helpful when designing switching regulators, since considerable power is lost while the switching element is traversing its active region. Figure 7-28 is the schematic of a 50-W, 200-kHz regulator using the RF231, a 9-A, 150-V DMOS device. The regulator output is 5 V at 10 A with ripple at less than 100 mV p-p. No output current limiting is included, although it may be added. Input supply is 28 V dc.

The 710 comparator acts as an oscillator, using L_1 as a reactive element and R_8 for hysteresis. C_5 couples the output ripple to the negative input of the comparator, where it is rejected as a common-mode signal. D_4, R_4, R_5, and C_3 form a bootstrap circuit which drives the gate 15 V more positive than the 28-V input. Six paralleled capacitors filter the output. The total impedance of one capacitor at 200 kHz is 0.05 Ω, and 0.01 Ω is needed to filter the 10-A peak-to-peak ripple current. Q_3 is the heart of a soft-startup circuit.

Operation at 200 kHz, rather than the usual 20 to 25 kHz, has several advantages:

1. A smaller inductor, with lower dc (copper) losses, is used.

2. A smaller filter capacitor may be used.

3. The regulator responds faster to sudden changes in the load.

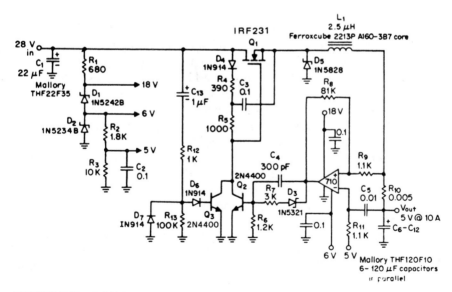

FIGURE 7-28   A 200-kHz switching regulator.

TABLE 7-1 A COMPARISON OF SWITCHING REGULATORS (28 V IN, 5 V AT 10 A OUT)

	20 kHz		200 kHz	
	Bipolar	*DMOS*	*Bipolar*	*DMOS*
Efficiency	82%	79%	72%	75%
Power output	50.0	50.0	50.0	50.0
Total input power	60.7	63.6	69.4	67.0
Fixed losses			4.85 W	
Drive power	0.17 W	0.44 W	1.4 W	0.87 W
Switching losses	1.9	0.55	9.6	3.7
Saturation losses	3.2	7.2	3.2	7.2
AC core losses		0.06		0.2
DC coil losses		0.49		0.13
Approximate recovery time for a 40% change in load	100 μs		10 μs	
Inductor core	3019 pot core ~0.85 in³, 1.2 oz		2213 pot core ~0.31 in³, 0.43 oz	
Capacitors	8 × 220 μF 1.0 in³		6 × 120 μF 0.45 in³	

High-frequency operation does reduce the overall efficiency somewhat, because of switching losses, but not as much as it would in a comparable design using a bipolar or bipolar darlington transistor as the switching element. Table 7-1 compares bipolar and DMOS regulators operating at 20 kHz and 200 kHz. A circuit similar to Fig. 7-28 is assumed.

The high input impedance and linear transfer characteristic of DMOS make it easy to control either the average or the surge current to a load. Figure 7-29 is a simple light dimmer circuit which varies the average current into the light bulb by controlling the saturation current of the DMOS. R_1 and R_2 make the control of brightness more linear with the potentiometer shaft rotation. The disadvantage of this circuit is that the 2N6661 operates in its linear region, so considerable power is wasted when the light is dimmed.

A more efficient method of varying the average current to the load is with pulse-width modulation (Fig. 7-30). The 4011 oscillates with a duty cycle which is determined by the ratio of R_1 and R_2, and drives the DMOS with 12-V pulses. Since the 2N6661 is either fully on or fully off, very little power is dissipated in the regulator itself.

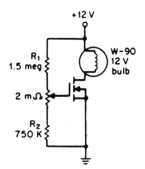

FIGURE 7-29   A linear light dimmer circuit. Not recommended except for very low wattage lamps.

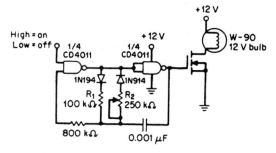

FIGURE 7-30   An efficient light dimmer circuit that dramatically limits the power dissipation of the power DMOSFET.

A similar circuit can be used as an inexpensive audio alarm system. The 4011 gate provide a 2-kHz square wave to the 2N6661, which directly drives an 8-Ω speaker (Fig. 7-31).

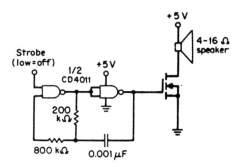

FIGURE 7-31   A 2-kHz audio alarm.

7-6 DRIVE CONSIDERATIONS

Many loads—motors and incandescent light bulbs, for instance—have an undesirably low impedance, resulting in high surge currents when power is first applied. The soft-startup circuits in Fig. 7-32 and 7-33 will minimize or eliminate these current surges, which in the case of an incandescent light bulb will increase life considerably. Adjust the 1-MΩ potentiometer in Fig. 7-32 until the desired maximum current is obtained, or use a fixed divider if a wider tolerance is allowable. R_1 and C_1 in Fig. 7-33 have a 0.1-s time constant to increase the drive voltage, and hence the drain current, of the 2N6661 gradually.

While the 2N6661 can drive nearly any load of 0.9 A or less (3 A under pulsed conditions), the gate must be driven with a high enough enhancement voltage to support the required current. Refer back to Fig. 7-5: When the DMOS is driven by TTL providing a maximum V_{GS} of 5 V, the saturated drain current is 800 mA. If a minimum drain current of 0.9 A is required, a worst-case minimum of 6 V must be applied to the gate.

Applying more than the minimum enhancement voltage—15 V rather

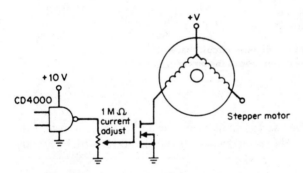

FIGURE 7-32 A circuit which current-limits the drive to the motor.

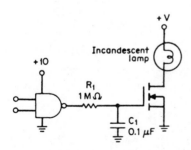

FIGURE 7-33 A soft start-up circuit reduces the cold current of the bulb.

than 10, for instance—has two desirable effects: the ON resistance is reduced, and an extra margin of safety is provided to allow for the decrease in drain current as the 2N6661 heats up. It is possible, as the drain current decreases with temperature, for the 2N6661 to actually come out of saturation, which further increases dissipation.

7-7 TEMPERATURE CONSIDERATIONS

Typically, the saturation drain current of DMOS decreases 0.5 to 0.6%/°C due to the decrease in the mobility of electrons in silicon as temperature increases. The drain-to-source ON resistance, $r_{DS(on)}$, increases exponentially with junction temperature and, consequently, so does the power dissipated by the MOSFET for a given rms drain current. If we assume a worst-case situation of 0.6%/°C decrease in drain current, the ON resistance at a given temperature can be expressed in terms of the resistance at the ambient temperature [r_{DS} (T_A)] by the expression,

$$r_{DS}(T) = r_{DS} (T_A)e^{0.006} dT \qquad (7\text{-}2)$$

where $dT = (T - T_A)$, the rise in temperature.

To better understand the problems of operation under load where the DMOS heats up, we need to consider both the steady-state thermal model and the transient, or switching, thermal model.

7-7-1 The Steady-State Thermal Model

Possibly the simplist steady-state thermal model is that shown in Fig. 7-34. From this model we can write an expression for junction tempera-

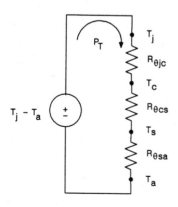

FIGURE 7-34 Steady-state thermal model. (Used with permission from *Siliconix Applications Handbook*, copyright Siliconix Incorporated, 1984.)

ture, T_j:

$$T_j = T_A + (Rth_{jc} + Rth_{cs} + Rth_{sa})P_T \qquad (7\text{-}3)$$

That expression enclosed within the parenthesis may be expressed as Rth_{ja}—the thermal resistance between the junction and ambient. Equation (7-3) simplifies to:

$$T_j = T_A + Rth_{ja}\, P_T \qquad (7\text{-}4)$$

Despite the apparent simplicity of Eq. (7-4), the term P_T is an exponential function of T_j which lends some difficulty in the execution of the equation.

The elements of these equations not previous defined are:

1. T_A is generally the ambient operating temperature of the environment, often set at 25°C. However, it can be any reasonable temperature up to but not exceeding the maximum operating temperature of the transistor.

2. Rth_{jc}—the thermal resistance, junction-to-case depends upon several parameters, specifically the die size, the method of attachment to the case and, of course, the type of case selected. The same transistor die packaged in a TO204 will have a lower thermal resistance than when packaged in a TO205.

3. Rth_{cs}—the thermal resistance, case to (heat)sink is determined both by the selected package (viz., TO204 or TO205) and how we attach the package to the heat-sink.

4. Rth_{sa}—the thermal resistance, heat sink to ambient is controlled exclusively by the design of the heat sink itself.

5. P_T is the total power dissipated by the DMOS FET.

7-7-2 The Transient Thermal Model

Quite often the DMOS is used in pulsed applications as we have seen in the many figures in this chapter leading up to this section. For these applications we must modify the steady-state model to account for the thermal capacity of each element: the die, the case, and the heat sink. The thermal model resembles the steady-state model but with the addition of capacitors, as shown in Fig. 7-35. These capacitors represent the thermal mass of each component.

From the transient thermal model the transient thermal impedance plot, shown in Fig. 7-36, has become the basis for most designs from which the safe operating area, shown in Fig 7-37, may be derived.

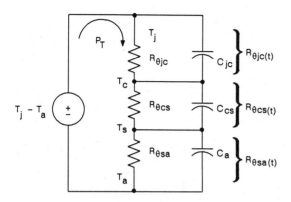

FIGURE 7-35 Steady-state thermal model amended to show the effects of mass. (Used with permission from *Siliconix Applications Handbook,* copyright Siliconix Incorporated, 1984.)

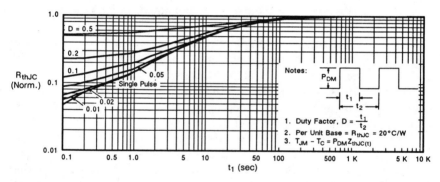

FIGURE 7-36 Transient thermal response time is a function of packaging, die size, and pulse duration. (Used with permission from *Siliconix Applications Handbook,* copyright Siliconix Incorporated, 1984.)

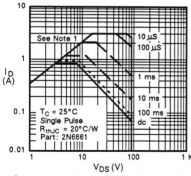

[1] Operation in this area may be limited by $r_{DS(ON)}$

FIGURE 7-37 Safe operating area 2N6661. (Courtesy of Siliconix Incorporated.)

7-8   PARALLEL AND SERIES OPERATION

If the required current exceeds the capability of one device, then several devices may be paralleled. Paralleling power DMOS can prove disastrous in some situations if precautions are not observed. For the low-power DMOS FETs paralleling is generally not a problem; the probems arises when we try paralleling high-power devices. The varied situations where we need to exercise precaution may be simply divided into *steady-state* and *switching*.

When operating in the linear mode (steady-state), one precaution concerns itself with how hard we are driving the DMOS. Operating a reasonably high-power DMOS FET in the linear mode generally finds a *negative* temperature coefficient of $r_{DS(on)}$. Consequently, in a group of parallel devices if one conducts more current than its neighbors, its temperature will rise, forcing the ON resistance down, leading to increased current hogging. The result is thermal runaway and eventual destruction of the device. The solution is to use "emitter ballast," viz., small source resistors to provide a small, but effective, amount of degeneration.

Because of the inherently high gain, adding a small amount of series resistance in the gate is recommended. Usually 50 Ω is adequate.

When using power DMOS in switching applications, paralleling may not be a problem provided the switching action is swift. Possibly a more insidious problem is layout: keeping parasitic series inductance and shunting capacitance to an absolute minimum.

Devices may be connected in series to increase breakdown voltage, as shown in Fig. 7-38. R_1 and R_2 are larger because the drive current to the gate of Q_2 is small, while C_1 and C_2 form a capacitive divider which dynamically balances the gate drive and also ensures fast switching times by coupling charge to the gate of Q_2. C_1/C_2 should be approximately equal to R_2/R_1, with allowance for stray capacitance and the enhancement voltage of Q_2. The bottom of the divider chain is returned to

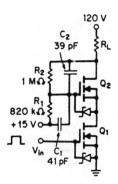

FIGURE 7-38   Breakdown voltage may be doubled by connecting two devices in series.

$+15$ V, rather than ground, to ensure sufficient enhancement for Q_2 when the devices are ON. If resistor and capacitor values are properly selected, any number of DMOS may be series-connected in this manner.

7-9 AMPLIFIER APPLICATIONS

The constant-g_{fs} region of DMOS makes it well suited for linear applications. Distortion is low over a wide dynamic range when properly biased. Figure 7-39 is a plot of the harmonic distortion vs. output voltage for a simple class A test circuit employing the 2N6661 DMOS. Distortion rises almost linearly with output voltage at low signal levels, but then rises more sharply as the positive signal peaks extend into the nonlinear g_{fs} region and the negative peaks saturate the device. The voltage gain of the circuit is about 6.5, equal to $g_{fs}R_L$ (0.27 mho $\times$ 24Ω).

Figure 7-40 shows that the frequency response of a simple class A stage

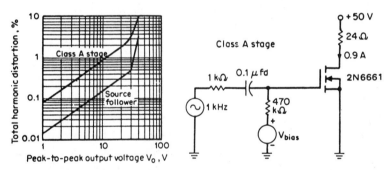

FIGURE 7-39 Harmonic distortion vs. voltage output for a simple class A stage and a source follower.

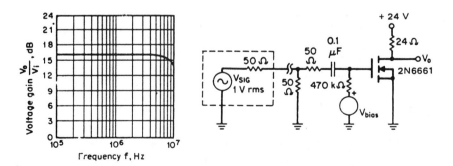

FIGURE 7-40 Frequency response of a simple class A stage.

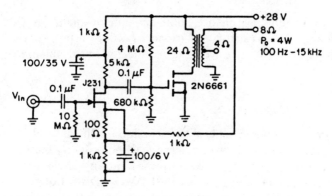

FIGURE 7-41   A simple audio power amplifier.

is flat to almost 10 MHz. The simple audio amplifier shown in Fig. 7-41 is equivalent to the audio output stage of many inexpensive radio and television receivers and phonographs. Power output is about 2 W from 100 Hz to 15 kHz. The design is greatly simplified by the use of an output transformer, and overall distortion is kept relatively low (2 percent at 1 W) by 10 dB of negative feedback. No thermal stabilization components are needed, since the drain current has a negative temperature coefficient.

7-10   SUMMARY

Power FETs typically rely on a short channel length and vertical current flow to increase current density and power capability. Their outstanding features, compared to bipolar transistors, include negligible dc drive current, extremely fast switching times, no minority-carrier storage time, a complete lack of secondary breakdown, and low distortion. They are being designed into numerous power applications, including both general-purpose and high-speed switchers, and switching regulators.

BIBLIOGRAPHY

Baliga, B. Jayant and Dan Chen (ed.): *Device Design and Applications,* IEEE Press, The Institute of Electrical and Electronic Engineers, New York, 1984.

Oxner, Edwin S.: *Power FETs and Their Applications,* Prentice-Hall, Englewood Cliffs, NJ, 1982.

Oxner, Edwin S.: *FET Technology and Application,* Marcel Dekker, New York, 1989.

Siliconix, Inc., *Siliconix MOSPOWER Applications Handbook,* Santa Clara, CA, 1984.

8

FETs IN INTEGRATED
CIRCUITS

8-1 INTRODUCTION

Field-effect transistors are now a basic component in many integrated circuits, both analog and digital. While any type of FET—n- or p-channel, MOS or junction—can be fabricated on the same substrate with either devices of the same type or nearly any other kind of FET or bipolar device, some combinations are naturally more popular and useful than others. In this chapter we shall describe some of the more popular FET and FET-bipolar integrated-circuit basic processes, although many variations on each process have been developed to suit particular applications and manufacturing capabilities.

8-2 MOSFET PROCESSES

Figure 8-1 is an all-PMOS process, the earliest and simplest MOS process. Only five masking steps are needed (p^+ diffusion, gate oxide, contact, metal, and oxide), so the cost is low and the yield high. Furthermore, the packing density (number of MOSFETs per unit area) is much higher than that of a bipolar process because no isolation diffusions are needed between devices. A high packing density reduces the cost of a function. The PMOS process is used primarily for medium- to high-complexity digital circuits, but has recently lost favor to NMOS.

NMOS, shown in Fig. 8-2, not only offers the same advantages of simplicity, low cost, and high packing density as PMOS, but it has a

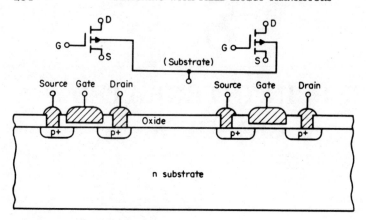

FIGURE 8-1   The PMOS process.

better speed-power product because negative charge carriers (electrons) have a higher mobility in silicon than do positive charge carriers (holes). This process was developed later than PMOS because it is much more sensitive to ionic contamination in the gate oxide, and only recently have processes been clean enough to ensure a high yield. However, NMOS has undergone extensive technical development and is a low-cost and widely used process for digital integrated circuits such as micro-processors and memory.

CMOS (complementary MOS), Fig. 8-3, is a combination of both n- and p-type MOSFETs. CMOS digital circuits dissipate very little power in the quiescent state because either the p or the n MOSFETs—but never both—are ON within a logic element, so there is no current flow except for a minimal leakage. At high switching frequencies (above sev-

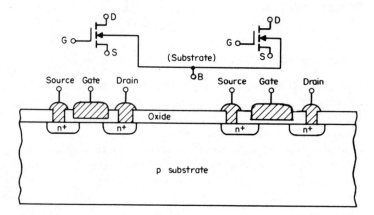

FIGURE 8-2   The NMOS process.

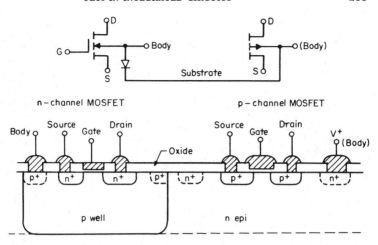

FIGURE 8-3   The CMOS process.

eral hundred kilohertz), however, the power dissipation becomes comparable to NMOS because circuit capacitances must be charged and discharged—which requires additional current. CMOS also requires more chip area for a given logic function because the *n*-channel MOSFET must be diffused into an isolated *p*-type well; this requires at least two more masking steps than PMOS or NMOS and is therefore more costly and difficult to fabricate. For digital circuits, CMOS is used when the complexity is low to medium; highly complex CMOS circuits are designed only when extremely low-power operation at low frequencies is needed—wristwatch circuits, for example.

CMOS is not confined purely to digital circuits, however—it is very popular for analog switches, operational amplifiers, and systems requiring both analog and digital circuitry. In an analog switch, the parallel combination of *p*- and *n*-channel MOSFETs exhibits a nearly constant

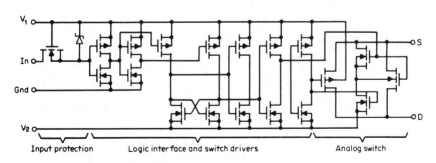

FIGURE 8-4   A CMOS analog switch.

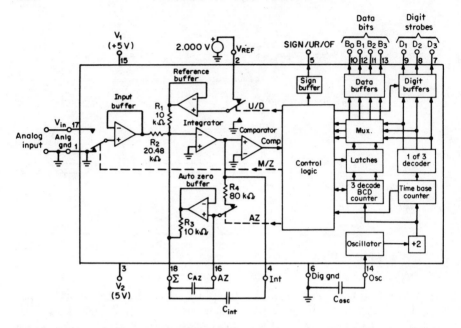

FIGURE 8-5   A monolithic CMOS chip containing both analog and digital circuitry.

resistance to any analog voltage between the positive and negative supply voltages. In a CMOS operational amplifier, the output is capable of excursions to within several millivolts of either supply rail versus several hundred millivolts or more for a more conventional bipolar design; a CMOS op amp also has several orders of magnitude lower input bias current than does a bipolar op amp, and both greater linearity and greater dynamic range than amplifiers made exclusively with either PMOS or NMOS. Figure 8-4 shows a CMOS analog switch system which

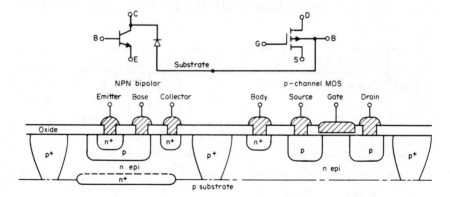

FIGURE 8-6   The bipolar-PMOS process.

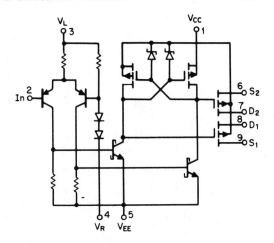

FIGURE 8-7 A bipolar-PMOS analog switch.

contains a digital interface and driver circuitry in addition to the analog transmission gate.

Figure 8-5 is a block diagram of the LD130, a CMOS three-digit analog-to-digital converter. Notice that the buffers, integrator, and comparator are analog circuits (operational amplifiers), while the control logic and circuitry to the right of the logic is all digital. The analog switch functions utilize MOSFETs.

8-3 BIPOLAR-FET COMBINATIONS

The bipolar-PMOS process (Fig. 8-6) is actually a forerunner of the previous MOSFET process; it was commercially developed in 1968 for

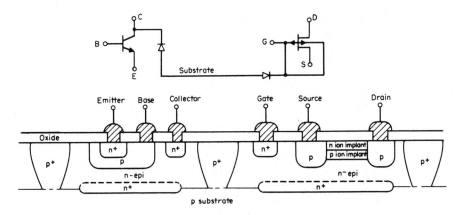

FIGURE 8-8 The BIFET (bipolar-JFET) process.

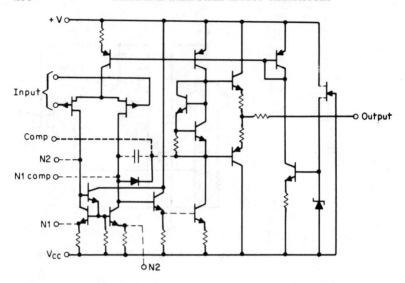

FIGURE 8-9   A BIFET op amp.

fabricating monolithic analog switch driver-gate combinations. The only significant difference between this process and the standard planar bipolar process is an extra masking step for the PMOS gate, so the cost and complexity are only slightly greater. The bipolar-PMOS process is used in a variety of analog circuits—analog switches (Fig. 8-7), A-to-D converters which require MOSFET-input operational amplifiers and analog signal switching, and smoke-detector ICs which require a MOSFET-input comparator to interface with an ion-chamber smoke sensor.

Figure 8-8 is a versatile development—the BIFET (bipolar-JFET) process. Like the bipolar-PMOS process, it is basically a planar bipolar

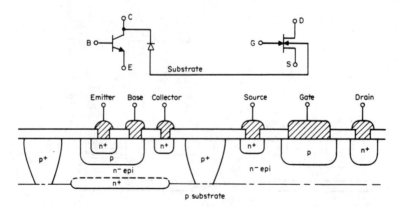

FIGURE 8-10   An n-channel JFET compatible with standard bipolar processing.

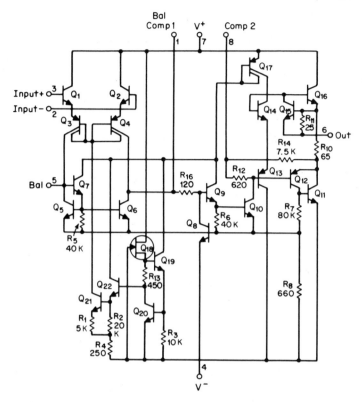

FIGURE 8-11 An integrated circuit operational amplifier using a JFET (Q_{18}) to ensure proper startup.

process, but with extra ion-implant processing steps to fabricate the channel of the p-channel JFET. This process is also used for a number of analog applications, including FET-input operational amplifiers (Fig. 8-9) and JFET analog switch/driver combinations.

The standard planar bipolar process can, with no extra processing steps, produce JFETs, but their parameters are difficult to control and close matching is nearly impossible. The process is shown in Fig. 8-10. JFETs of this type are used primarily for noncritical biasing and current sources in analog integrated circuits, especially since they are always ON when power is first applied and, therefore, ensure the startup of bipolar bias circuits (Fig. 8-11).

BIBLIOGRAPHY

Carr, W., and J. Mize: "MOS/LSI Design and Application," McGraw-Hill, New York, 1972.

Landsburg, G. F.: "A Charge Balancing Monolithic A/D Converter," *Proceedings of the 1977 IEEE International Solid-State Circuits Conference,* pp. 98, 99.

Penney, W., and L. Lau: "MOS Integrated Circuits," Van Nostrand Reinhold Company, New York, 1972.

Index